MW01007601

Life of Fred®

Mineshaft

Life of Fred®
Mineshaft

Stanley F. Schmidt, Ph.D.

Polka Dot Publishing

ISBN: 978-1-937032-08-1

Library of Congress Control Number: 2012950880
Printed and bound in the United States of America

Polka Dot Publishing Reno, Nevada

Order again from:
JOY Center of Learning
http://www.LifeofFredMath.com

Questions or comments? Email the author at lifeoffred@yahoo.com

Third printing

for Goodness' sake

or as J.S. Bach—who was
never noted for his plain
English—often expressed it:

Ad Majorem Dei Gloriam

(to the greater glory of God)

If you happen to spot an error that the author, the publisher, and the printer missed, please let us know with an email to: lifeoffred@yahoo.com

As a reward, we'll email back to you a list of all the corrections that readers have reported.

A Note Before We Begin
Life of Fred: Mineshaft

There are several keys to success to learning math,
to being a good cook,
to doing tricks on a skateboard,
to writing applications for a cell phone,
to climbing mountains, and
to playing hide-and-seek with a tiger.

One important key is . . .

Effort.

The most common mistake is to put in too little effort. Human beings often like to take the easy road.

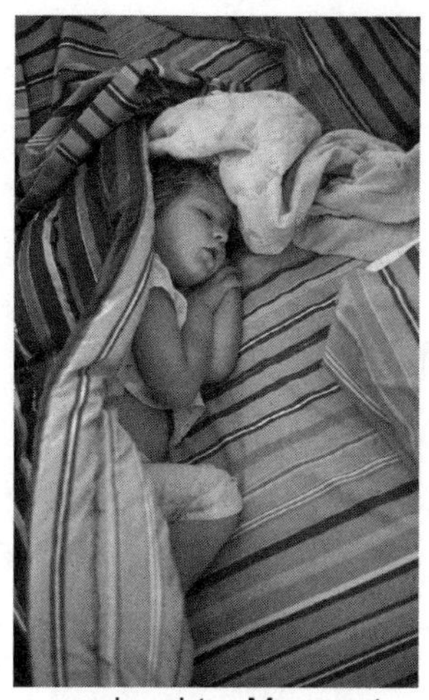
my daughter Margaret

There is nothing wrong with resting up a little bit before you tackle the main work of the day.

There is nothing wrong with heading off to sleep after running around hard all day.

But it is wrong if your life consists of just sleeping, fooling around, napping, goofing off, dozing, eating, snoozing, and resting,
. . . unless you were born yesterday.

In this book, please make the effort of doing the *Your Turn to Play* problems before you turn the page and look at my answers.

Write down the answers, even though you can do many of the problems in your head. You will learn more that way.

The second mistake with regard to effort is the opposite of the first. Instead of putting in too little effort, you put in too much effort. One student in a thousand makes this error.

Sometimes when you get to a mountain of granite, it is better to walk around it than try to tear your way through it.

Hint: *Life of Fred: Mineshaft* is not a mountain of granite.

Margaret's silly father tearing the mountain apart

CALCULATORS?

Not now. There will be plenty of time later after finishing *Life of Fred: Fractions* and *Life of Fred: Decimals and Percents.*

Right now in arithmetic, our job is to learn the addition and multiplication facts by heart.

Contents

Chapter One
The Prize

Kingie had won the university seal contest. That seal would now be used on all the official documents of KITTENS University. Every diploma, every transcript, and every letter would have . . .

THE KITTEN CABOODLE newspaper had advertised that the winner of the contest would "win big," but had not said what the prize would be.

Kingie didn't care. He was much more interested in painting than getting prizes or making money. As long as he had oil paints and a quiet place to work, he was happy.

The staff at THE KITTEN CABOODLE didn't know that Kingie was not concerned about a prize. If they had known, life would have been a lot easier for everyone.

The editor of the newspaper called a meeting of his staff. He asked them, "What prize should we give Kingie?"

Caboodle Editor

Ashley said, "Most people would be happy getting money."

Chris responded, "But we don't have a lot of money. Our newspaper only sells for a dime."

Drew asked, "What do we know about this guy Kingie? All we know right now is that he is a terrific artist. Let's get him something he would like."

Joyce got out a piece of paper and wrote . . .

Dear Kingie,

You are the winner of the university seal contest.

We here at The KITTEN Caboodle are wondering what prize you would like. We will try to get it for you—as long as it's not too expensive.

Sincerely,

Joyce

Joyce sent it by campus mail. Kingie received it two seconds later.

When Kingie read the letter, he didn't know how to respond. On the top floor of Kingie's fort were 16 safes (9 on the left wall and 7 on the right wall), and they were all filled with money.

Kingie's art had sold very well. If he received more money, he would just have to buy more safes.

Chapter One   The Prize

Kingie wrote back in his beautiful artistic handwriting . .

•

Dear Joyce,

Thank you for kind thoughts. I really do not need anything for myself. May I make a suggestion? There are a lot of hungry people in the world.

Would you please send one grain of rice to Afghanistan? If you use campus mail, it will not cost anything for shipping.

And two grains of rice to Akrotiri. And four grains of rice to Albania.

I have enclosed a list of places from K's Afternoon Dining menu. Each will receive twice as many grains as the previous place.

Love,
Kingie

Afghanistan 1 grain
Akrotiri 2 grains
Albania 4 grains
Algeria 8 grains
American Samoa 16
Andorra 32
Angola 64
Anguilla 128
Argentina 256
Armenia
Aruba
Australia
Austria
Azerbaijan
Bahamas
The Bahrain
Bangladesh
Barbados
Bassas da India
Belarus
Belgium
Belize
Benin
Bermuda
Bhutan
Bolivia
Bosnia and Herzegovina
Botswana
Bouvet Island
Brazil
British Virgin Islands
Brunei
Bulgaria
Burkina Faso
Burma
Burundi
Cambodia
Cameroon
Canada
Cape Verde
Central African Republic
Chad
Chile
China
Christmas Island
Clipperton Island
Cocos Colombia
Comoros
Congo
Cook Islands
Coral Sea Islands
Costa Rica
Cote d'Ivoire
Croatia
Cuba
Cyprus
Czech Republic
Denmark
Dhekelia
Djibouti
Dominica
Dominican Republic
Ecuador
Egypt
El Salvador
Equatorial Guinea
Eritrea
Estonia
Ethiopia
Europa Island
Falkland
Fiji
Finland
France
French Guiana
French Polynesia
Gabon
Gambia
Georgia
Germany
Ghana
Gibraltar
Glorioso Islands
Greece
Greenland
Grenada
Guadeloupe
Guam
Guatemala
Guernsey
Guinea
Guyana
Haiti
Honduras
Hong Kong
Hungary
Iceland
India
Indonesia
Iran
Iraq
Ireland
Isle of Man
Israel
Italy
Jamaica
Jan Mayen
Japan
Jersey
Jordan
Juan de Nova Island
Kazakhstan
Kenya
Kiribati
Korea North
Korea South
Kuwait
Kyrgyzstan
Laos
Latvia
Lebanon
Liberia
Libya
Liechtenstein
Lithuania
Luxembourg
Macau
Macedonia
Madagascar
Malawi
Malaysia
Maldives
Mali
Malta
Marshall Islands
Martinique
Mauritania
Mauritius
Mayotte
Mexico
Micronesia
Moldova
Monaco
Mongolia
Montserrat
Morocco
Mozambique
Namibia
Nauru
Navassa Island
Nepal
Netherlands
Netherlands Antilles
(We'll stop here.)

The staff at the newspaper was delighted with Kingie's request.

Ashley said, "It's neat that Kingie didn't demand a big prize for himself. Lots of people just want more, more, more, for themselves and are never satisfied."

Chris added, "And he was thinking of other people. I would like to meet this guy someday."

Drew got a dollar from the editor and headed out to buy a one-pound box of rice. Joyce started addressing the envelopes so that they could mail the rice to each country.

When Drew got back from the store, he opened the box and carefully poured it out onto a desktop. Ashley put one grain of rice in the Afghanistan envelope. Chris put two grains in the Akrotiri envelope. Drew put four grains in the Albania envelope.

Joyce kept addressing the envelopes. There were roughly 30 countries in each column, and there were 5 columns. ⇨ 150 countries.

Ashley told Drew, "Don't throw away the empty rice box. After we're done, we can put the leftover rice back in the box."

They put 8 grains into the Algeria envelope.
And 16 grains into the American Samoa envelope.
And 32 grains into the Andorra envelope.
And 64 grains into the Angola envelope.
And 128 grains into the Anguilla envelope.
And 256 grains into the Argentina envelope.
And 512 grains into the Armenia envelope.
And 1,024 grains into the Aruba envelope.

Your Turn to Play

1. They put 1,024 grains into the Aruba envelope.
Compute the next three envelopes:
They put ___?___ grains into the Australia envelope.
They put ___?___ grains into the Austria envelope.
They put ___?___ grains into the Azerbaijan envelope.

2. One grain of rice has a mass of 25 mg.

Wait! Stop! I, your reader, have a question. What does "mg" mean?

In the metric system, which most of the world uses, mg stands for milligram.

And what in the world is a milligram?

A thousand milligrams equals a gram. 1000 mg = 1 g. And a gram is roughly the weight of a raisin.

Thank you.

So how many grains of rice would weigh the same as a raisin?

Translation: How many grains of rice in a gram?

Translation: How many grains of rice in 1000 mg?

Translation: How many 25 mg are in 1000 mg?

Translation: Divide 25 into 1000.

3. There are about 454 grams in a pound.
How many grains of rice in a pound?

Translation: One pound equals how many grains of rice?

Translation: Given your answer to question 2 (which is how many grains of rice in a gram) and given that there are 454 grams in a pound, how many grains of rice in a pound?

Translation: 40×454 equals what?

.......COMPLETE SOLUTIONS.......

1. 2,048 for Australia, 4,096 for Austria, 8,192 for Azerbaijan.

2. To find out how many 25s are in 1,000, we divide 25 into 1,000.

$$\begin{array}{r} 40 \\ 25\overline{)1000} \\ \underline{100} \\ 00 \\ \underline{00} \end{array}$$

There are 40 grains of rice in a gram.

3. There are 40 grains of rice in a gram and 454 grams in a pound. If you don't know whether to add, subtract, multiply, or divide, the general rule is: restate the problem using easy numbers and notice which of the four operations you used. Suppose there are 3 grains of rice in a gram and 2 grams in a pound. Then there would be 6 grains of rice in a pound. You multiplied.

$$\begin{array}{r} 454 \\ \times\ 40 \\ \hline 0 \\ \underline{1816\ } \\ 18160 \end{array}$$

There are 18,160 grains of rice in a pound.

A grain of rice weighs *roughly* 25 mg.
There are *exactly* 1000 milligrams in a gram.
There are *roughly* 454 grams in a pound.

So there are *roughly* 18,160 grains of rice in a pound.

We can round 18,160 off to 18,000.
18,160 is closer to 18,000 than it is to 19,000.

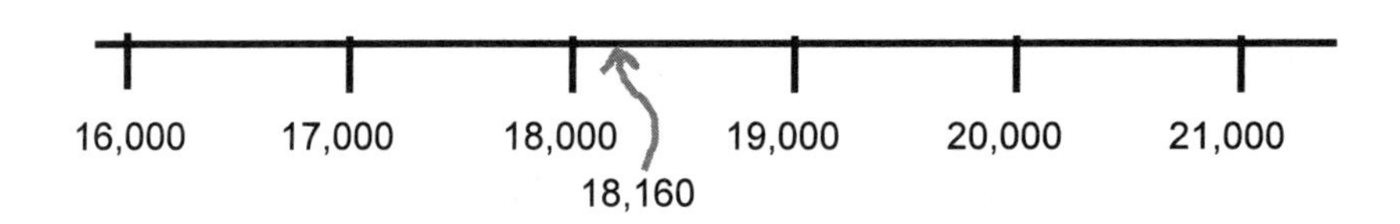

Chapter Two
Count the Cost

Drew took the empty box of rice and wrote *18,000 grains* on it. When Chris put 8,192 grains of rice into the Azerbaijan envelope, everyone suddenly realized that the 16,384 grains of rice that they would be sending to Bahamas* would be almost a whole box of rice.

Drew got the editor's charge card and headed to a local warehouse store and bought three 25-pound sacks of rice. He figured that 75 pounds of rice should be enough.

Afghanistan 1 grain
Akrotiri 2 grains
Albania 4 grains
Algeria 8 grains
American Samoa 16
Andorra 32
Angola 64
Anguilla 128
Argentina 256
Armenia
Aruba
Australia
Austria
Azerbaijan
Bahamas *one pound of rice*
The Bahrain
Bangladesh
Barbados
Bassas da India
Belarus
Belgium
Belize
Benin
Bermuda
Bhutan
Bolivia

While Drew was at the store, Joyce got a pencil and started computing. The Bahrain would get 2 pounds of rice.

---

* Its official name is Commonwealth of The Bahamas.

Bangladesh would get 4 pounds. Barbados would get 8 pounds.

Bahamas *one pound of rice*
The Bahrain *2 pounds of rice*
Bangladesh *4 lbs.*
Barbados *8 lbs.*
Bassas da India *16 lbs.*
Belarus *32 lbs.*
Belgium *64 lbs.*
Belize *128 lbs.*
Benin *256 lbs.*
Bermuda *512 lbs.*
Bhutan *1024 lbs.*
Bolivia *2048 lbs.* *That's over a ton!* 2000 pounds = one ton

When Drew got back with the three sacks of rice, Joyce told him that 75 pounds would not be enough.

Drew started to head out the door to buy more rice at the warehouse store. Ashley, Chris, and Joyce all yelled, "Stop!" It was time to count the cost before they went any further.

Bermuda 512 lbs.
Bhutan 1024 lbs.
Bolivia one ton
Bosnia and Herzegovina 2 tons
Botswana 4 tons
Bouvet Island 8 tons
Brazil 16 tons
British Virgin Islands 32 tons
Brunei 64 tons
Bulgaria 128 tons
Burkina Faso 256 tons
Burma 512 tons
Burundi 1024 tons That's 1,024 × 2,000 = 2,048,000 which is about two million pounds
Cambodia 4 million pounds
Cameroon 8 million pounds
Canada 16 million pounds
Cape Verde 32 million pounds
Central African Republic 64 million pounds
Chad 128 million pounds
Chile 256 million pounds
China 512 million pounds
Christmas Island 1024 million pounds A thousand million is a billion. 1,000,000,000
Clipperton Island 2 billion pounds
Cocos Colombia 4 billion pounds
Comoros 8 billion pounds
Congo 16 billion pounds
Cook Islands 32 billion pounds
Coral Sea Islands 64 billion pounds
Costa Rica 128 billion pounds
Cote d'Ivoire 256 billion pounds
Croatia 512 billion pounds
Cuba 1024 billion pounds A thousand billion is a trillion. 1,000,000,000,000
Cyprus 2 trillion pounds
Czech Republic 4 trillion pounds
Denmark 8 trillion pounds
Dhekelia 16 trillion pounds
Djibouti 32 trillion pounds
Dominica 64 trillion pounds

Drew told the editor that they needed to send 1024 billion pounds of rice to Cuba.

The editor said, "That's a lot of rice. I don't think we can do that." He checked on his computer and found that the entire global production of rice in 2010 was 1023 billion pounds.

Here is Kingie's original list.

Afghanistan 1 grain
Akrotiri 2 grains
Albania 4 grains
Algeria 8 grains
American Samoa 16
Andorra 32
Angola 64
Anguilla 128
Argentina 256
Armenia
Aruba
Australia
Austria
Azerbaijan
Bahamas
The Bahrain
Bangladesh
Barbados
Bassas da India
Belarus
Belgium
Belize
Benin
Bermuda
Bhutan
Bolivia ***one ton***
Bosnia and Herzegovina
Botswana
Bouvet Island
Brazil
British Virgin Islands
Brunei
Bulgaria
Burkina Faso
Burma
Burundi
Cambodia
Cameroon
Canada
Cape Verde
Central African Republic
Chad
Chile
China
Christmas Island
Clipperton Island
Cocos Colombia
Comoros
Congo
Cook Islands
Coral Sea Islands
Costa Rica
Cote d'Ivoire
Croatia
Cuba ***one trillion pounds***
Cyprus
Czech Republic
Denmark
Dhekelia
Djibouti
Dominica
Dominican Republic
Ecuador
Egypt
El Salvador ✶
Equatorial Guinea
Eritrea
Estonia
Ethiopia
Europa Island
Falkland
Fiji
Finland
France
French Guiana ✶✶
French Polynesia
Gabon
Gambia
Georgia
Germany
Ghana
Gibraltar
Glorioso Islands
Greece
Greenland
Grenada
Guadeloupe
Guam
Guatemala
Guernsey
Guinea
Guyana
Haiti
Honduras
Hong Kong
Hungary
Iceland
India
Indonesia ✶✶✶
Iran
Iraq
Ireland
Isle of Man
Israel
Italy
Jamaica
Jan Mayen
Japan
Jersey
Jordan
Juan de Nova Island
Kazakhstan
Kenya
Kiribati
Korea North
Korea South
Kuwait
Kyrgyzstan
Laos
Latvia
Lebanon
Liberia
Libya
Liechtenstein
Lithuania
Luxembourg
Macau
Macedonia
Madagascar
Malawi
Malaysia
Maldives
Mali
Malta
Marshall Islands
Martinique
Mauritania
Mauritius ✶✶✶✶
Mayotte
Mexico
Micronesia
Moldova
Monaco
Mongolia
Montserrat
Morocco
Mozambique
Namibia
Nauru
Navassa Island
Nepal
Netherlands
Netherlands Antilles
(We'll stop here.)

A trillion = 1,000,000,000,000 which is twelve tens multiplied together. In algebra, this is represented as 10^{12}.

✶ By El Salvador, we hit one quadrillion pounds of rice. 10^{15}

✶✶ By French Guiana, one quintillion pounds of rice. 10^{18}

✶✶✶ By Indonesia, we are roughly at the weight of the sun, which is a little over 6 sextillion *tons*. 6 times 10^{21}

✶✶✶✶ There are 400 billion stars in a typical galaxy. After 39 more doubles, we get to Mauritius, to which we are shipping the weight of a galaxy of stars.

There are about 80 billion galaxies in the observable universe, so another 36 countries down from Mauritius, would bring us to the mass of the entire universe as we know it.

Drew handed the editor back his charge card and told him, “We won’t need it.”

Ashley, Chris, and Drew each picked up one of the 25-pound bags of rice. They carried them to the third floor of the Mathematics Building, down the hallway past the nine vending machines.

Chris knocked on the door and Fred answered it.

Drew asked, “Is this where Kingie lives?”

Fred nodded and they handed him the three bags of rice.

Actually, they tossed the bags to him. The bags weighed 75 pounds. Fred weighs 37 pounds. In symbols, 75 > 37. (Seventy-five is greater than thirty-seven.)

Kingie came out of his fort and asked what all the noise was. Fred was turning a little purple. He couldn’t breath very easily. All he could say was, “I think these are for you.” His voice was not very loud.

Kingie pushed the bags off of Fred.

Fred took several deep breaths.

Kingie said, “I guess they didn’t like my suggestion to feed the world.”

Fred didn’t know about Kingie’s letter to Joyce, and, therefore, he didn’t know what Kingie was talking about.

Kingie told Fred to store the rice in his desk “for later.” Fred thought that that was a good idea since he wasn’t hungry right now, and Kingie never ate anything.

Your Turn to Play

1. Suppose there is a country with a population of 19,416 and that three-eights ($\frac{3}{8}$) of the people are hungry. How many people is that?
Hint: Recall that $\frac{2}{5}$ of 35 means *35 times 2 and divide the result by 5.*

2. Is this true: one thousand > one million?

3. How much would twenty-five 25-pound bags of rice weigh?

4. Those three bags of rice weighed 75 pounds. Fred weighed 37 pounds. How much heavier were the bags than Fred?

5. A really silly question . . .

Is 75 pounds < 82 days? (< means "less than")

6. That was fun to invent.
Here's some more: Is a quart < a duck?
Is 45° < California?
Is a piece of pie < the letter *B*?
Is George Washington's birthday < silver?

Your turn. You invent a silly question.

.......COMPLETE SOLUTIONS.......

1. We want $\frac{3}{8}$ of 19,416.

   That means 19,416 times 3 and divide the result by 8.

$$\begin{array}{r} 19416 \\ \times \quad 3 \\ \hline 58248 \end{array} \qquad \begin{array}{r} 7281 \\ 8\overline{)5\,8^{2}2^{6}4\,8} \end{array}$$

using short division

There are 7,281 hungry people.

2. Is 1,000 > 1,000,000?

   Is 1,000 greater than 1,000,000?

No. It isn't. It is true that 1,000 < 1,000,000.

One thousand is less than one million.

3.
$$\begin{array}{r} 25 \\ \times 25 \\ \hline 125 \\ 50 \\ \hline 625 \end{array}$$

Twenty-five 25-pound bags weigh 625 pounds.

4.
$$\begin{array}{r} 75 \\ -\ 37 \\ \hline 38 \end{array}$$

The rice is 38 pounds heavier than Fred.

5. There is no way to compare pounds with days.

6. This is too much fun.

   Is an acre < a glass of orange juice?

   Is 16 cubic feet < Jeanette MacDonald?

   Is a shoe < a baseball?

   Is Shakespeare's *Hamlet* < Denmark?

Help! I can't stop thinking of more of these!

Chapter Three
A New Hobby

It was the middle of the afternoon. All the well-wishers had come to see Kingie. His prize had been delivered. It was now quiet. Really quiet. The university president had canceled classes until graduation day, and that was about a week from now.

Almost everyone had either gone home or gone on vacation. Fred had tried to go to Camp Horsey-Ducky but that didn't work out.

Kingie was happy to continue oil painting for the next week.

Fred didn't have any classes to teach. He didn't have any students to help. He sat down with a piece of paper and wrote out the things he could do each day for the next week.

Start with 24 hours in a day. I sleep for 9 hours each day since I'm five years old.

$$\begin{array}{r} 24 \\ -\ 9 \\ \hline 15 \end{array}$$

So I'm awake 15 hours per day.

I want to spend three hours each day reading. I really need that. I don't want to get super old--like 15 years old--and be ignorant.

$$\begin{array}{r} 15 \\ -\ 3 \\ \hline 12 \end{array}$$

I do an hour's worth of jogging each morning. 12 – 1 = 11 hours. I wasn't planning on doing a lot of eating this week. 11 – 0 = 11 hours.

I don't have any shopping to do. Neither Kingie nor I need anything right now. 11 – 0 = 11 hours.

Then Fred remembered a Valentine card that Darlene had sent. Joe had used the back of the card to do his math homework. Joe wasn't very interested in getting Valentine's cards. He didn't like what he called "mushy stuff."

The card gave Fred an idea: For the next week, I can study archery. I've never done that before. That should be fun.

Fred looked in the phonebook under "Archery."

There were seven archery stores all on the 500 block of Main Street. Fred chose the one with the biggest ad. He didn't read the fine print on the right side of the ad.

He grabbed his checkbook, said goodbye to Kingie, and raced down the hallway past the nine vending machines (four on one side and five on the other).

He headed down the two flights of stairs two at a time and counted 2, 4, 6, 8, 10, 12, 14, 16, 18, 20, 22, 24 and then realized that he had dropped his checkbook near the top of the stairs and headed back 22, 20, 18, 16, 14, 12, 10, 8, 6, picked up his checkbook and headed down the stairs 6, 8, 10, 12, 14, 16, 18, 20, 22, 24, 26, 28 and out into the open air.*

---

* Is this the longest sentence you have ever read?

Once, Fred had seen Alexander, who is six feet tall, climb the stairs three at a time, 3, 6, 9, 12, 15, 18, 21, 24, 27. Fred had tried to do that, but his legs were too short.

If Fred were a little bird, he could fly up the stairs and never touch any of the steps.

If Fred were a worm, he couldn't climb the stairs at all.

Wait! Stop! I, your reader, have a question.

Is this a good time to ask your question? Fred is ready to run to the Everybody's Archery store to start his archery hobby.

Yes. It's the perfect time. You just mentioned a Fred bird and a Fred worm. I need to know.

Know what?

Do Fred birds eat Fred worms?

That's a silly question.

Answer it! You are the author. You would know.

Have you ever seen Fred eat anything? A Fred worm is very safe around a Fred bird.

Thank you. I feel better now. You can get back to the story.

Fred ran through the KITTENS campus and down Main Street. In one block all the addresses were in the 100s, such as 104, 110, 113, 148, 155, 189.

All the addresses in the next block were in the 200s, such as 208, 227, 231, 277, 295.

All the addresses in the third . . . ***Stop! I, your reader, have another question. This is the only math book in the whole world in which the reader can stop the author and ask a question.***

What is your question?

Did you notice that all the addresses in the first block are in the 100s and in the second block are in the 200s?

Of course. I just said that.

No. That wasn't my question. I just noticed that this is the same way they number rooms in a tall building. The rooms in the 100s are on the first floor, and the rooms in the 200s are on the second floor. Did you ever notice that?

No, I didn't. Until you mentioned that, I had never made the connection between addresses on blocks and rooms on floors. Thank you.

If Fred had gone into Albert's or Bill's or Cindy's or Dale's or Flavio's or Glinda's archery shops, he could have bought a beautiful bow and some good arrows at a reasonable price. And he would have been treated nicely.

Instead, he went into Everybody's Archery, which is owned by C.C. Coalback.

Wait a minute! I hate to interrupt again, but I remember that Coalback and his sister were captured and sent off to jail back in Life of Fred: Kidneys. Fred got a $1,000 check for helping in the capture. How can they be running an archery store?

You are right. You have a good memory. They are still in jail, but that doesn't stop Coalback from owning a store. He hired someone to run it.

Her name was Methhunda.

She liked to wear black garbage sacks.

Your Turn to Play

1. If you sleep 8 hours out of every 24, then you sleep $\frac{8}{24}$ (eight twenty-fourths) of each day.

If you watch television for 6 hours each day, what fraction of the day is that?

2. Reduce the fraction that you got in question one.
Hint: To reduce $\frac{6}{9}$ you would divide top and bottom by 3.
$\frac{6}{9} = \frac{2}{3}$
Hint: To reduce $\frac{8}{12}$ you would divide top and bottom by 4.
$\frac{8}{12} = \frac{3}{4}$
Hint: To reduce $\frac{35}{60}$ you have to figure out what will divide evenly into both 35 and 60.

35 and 60 are not both even, so 2 won't work.

35 is not evenly divisible by 3, so 3 won't work.

To reduce $\frac{35}{60}$ you divide top and bottom by 5.

$\frac{35}{60} = \frac{7}{12}$

3. Here's counting by twos: 2, 4, 6, 8, 10, 12. . . .
Here's counting by threes: 3, 6, 9, 12, 15, 18, 21. . . .
Count by 4s up to 36.

4. If Fred eats zero dollars worth of food each day, how much would it cost to feed Fred for 3,987,052 days?

.......COMPLETE SOLUTIONS.......

1. If you watch television for 6 hours each day, that is $\frac{6}{24}$ (six twenty-fourths) of each day.

2. To reduce $\frac{6}{24}$ you divide top and bottom by 6 and get

$\frac{6}{24} = \frac{1}{4}$

Some readers might have started with $\frac{6}{24}$ and divided top and bottom by 3: $\frac{6}{24} = \frac{2}{8}$ and finished the problem by dividing top and bottom by 2: $\frac{2}{8} = \frac{1}{4}$

Some readers might have started with $\frac{6}{24}$ and divided top and bottom by 2: $\frac{6}{24} = \frac{3}{12}$ and finished the problem by dividing top and bottom by 3: $\frac{3}{12} = \frac{1}{4}$

> Three ways to do a problem! And they all get the same answer.

3. 4, 8, 12, 16, 20, 24, 28, 32, 36

4. Zero times any number always gives an answer of zero.
   $0 \times 3{,}987{,}052 = 0$

It's also true that $3{,}987{,}052 \times 0 = 0$.

Chapter Four
On Sale

When Coalback was looking for someone to run his Everyone's Archery store, he wanted someone who was just like himself.

He wanted someone

- ➤ who lied a lot
- ➤ who hated kids
- ➤ who liked to steal
- ➤ who couldn't be trusted.

Methhunda was perfect.

When Fred walked into the Everyone's Archery store, the first thing Methhunda said was, "Hey, kid! Do you got any money?"

Fred said, "I'm not carrying any cash with me."

Then she asked, "Do you have a credit card. You know, like Disastercard, Pisa card, or American Depression card."

Fred told her that he was too young to get a credit card.

"Then get out of here you brat!" she screamed.

Fred held up his checkbook and said that he had seven hundred dollars in his account.

Suddenly, she smiled and became nicey-nicey. "Well, my young man. You are most welcome to our store. I have a very nice bow-and-arrow set that was originally $3,599, but I can let you have it for $700."

Real Men Shoot Arrows

Methhunda pointed to the picture on the wall and told Fred, "Archery will make a man out of you. People look up to archers. Archery will build your muscles.

Archery is the coming new fad. Pretty soon everyone will be shooting arrows. You will be able to teach all your friends."

Methhunda didn't care whether what she was saying was true. All she was interested in is getting Fred's money.

✓ Archery isn't considered a body-building sport.

✓ People don't look up to you just because you own a bow and arrows.

✓ Archery isn't a new fad. People have been shooting arrows for centuries.

In short, Methhunda was a liar, just like Coalback.

She showed Fred the $700 archery set, the one that originally was $3,599.

"This is the finest archery set you could buy. [lie] You had better hurry before someone else comes and buys it. [lie] This set was handcrafted in Germany by the Bogenschütze family that has been in the archery business for generations. [lie]"

"The Bogenschütze family!" Fred was impressed. "*Bogenschütze* means archer in German," he said.

The truth was that Methhunda had made the set this morning. She found a twig on the ground and tied a string on it to make the bow. She took three sticks and sharpened one end of each in a pencil sharpener and glued feathers on the other end.

She stuck the set in a grocery sack, handed it to Fred, and said, "That will be $700." She didn't ask Fred whether he wanted to buy it or not.

Before Fred realized what was happening, he was writing out a check to Everybody's Archery for $700.

Two Stories at the Same Time

After Fred left the store, Methhunda turned off the lights and closed the store.

She took Fred's check to the KITTENS Bank and asked for cash.

The teller handed her $700.

She rushed to the bus station. The ticket agent asked her where she would like to go.

She said, "Anywhere. Give me a ticket on the first bus out of town."

Coalback had hired Methhunda for $6 per hour. He was surprised that she would work for so little money.

What he hadn't counted on was that she would steal from him.

Later, when he learned what Methhunda had done, his face turned red with anger. It was okay for him to steal from others but not for others to steal from him.

Fred loved his new archery set. He liked the fact that it was made by the Bogenschütze family in Germany.

He headed out to the Great Lawn on the KITTENS campus. Since almost everyone was on vacation, it would be a very safe place to shoot arrows.

When he opened the paper bag he noticed that one of the arrows was broken. (It actually was broken when Methhunda put it into the bag.)

When he pulled on the bow, it broke. *I must be getting really strong* he thought to himself.

The truth was that the bow was really weak.

Fred dropped the archery equipment into the garbage can. This was the end of Fred's archery hobby.

He was standing on the edge of the Great Lawn. It was 5 p.m. (p.m. means *post meridiem*, which is Latin for "after noon." p.m. means any time between noon and midnight.)

There you are. I've been looking all over for you. When I got to camp and you weren't on the hood of the car I wondered what had happened.

"Miss Ente!" Fred exclaimed. "I thought you had forgotten all about me."

"I didn't forget about you," she said. "I have searched for you for the last 24 hours or so. I just stopped here at the Great Lawn to munch some grass."

Fred asked, "Where's your doll, the one who was riding on you?"

"Oh, I left Dolly back at Camp Horsey-Ducky. I can go faster when I'm not carrying her."

Dolly

Intermission

It is amazing how fast humans can get used to something new.

When Fred first learned that Miss Ente could talk, he passed out.

Now, he could chat with this horse as if he had been talking to horses all his life.

Young people don't realize how much change old people have seen. When your author taught in college, no one owned a computer or a cell phone. No one.

Your Turn to Play

1. If the bow-and-arrow set was originally $3,599 and Fred bought it for $700, how much did he save?

2. The bow-and-arrow set was really worth $2. Fred bought it for $700. How much was he overcharged?

3. During a typical day Methhunda would say 292 things. Three-fourths ($\frac{3}{4}$) of her statements were lies. How many things would she say that were not true?

4. Miss Ente and Dolly, together, weigh 1482 pounds. Without Dolly, Miss Ente weighs 1457 pounds. How much does Dolly weigh?

5. 10^5 means $10 \times 10 \times 10 \times 10 \times 10$, which is 100,000.

7^2 means 7×7, which is 49. (The ^{2} is called an **exponent**.)

Compute each of these:

$9^2 = ?$

$5^2 = ?$

$2^5 = ?$

$1^{387} = ?$

What's more weird . . .

A) a horse that can talk or

B) using a little cell phone, which can fit in your shirt pocket, to talk with someone who lives half way around the world?

.......COMPLETE SOLUTIONS.......

1. $$\begin{array}{r} 3599 \\ -\ 700 \\ \hline 2899 \end{array}$$

He "saved" $2,899.

2. $$\begin{array}{r} \overset{6}{\cancel{7}}\,{}^{1}\overset{9}{\cancel{0}}\,{}^{1}0 \\ -\ \ \ 2 \\ \hline 698 \end{array}$$

He should have paid $2. He did pay $700.
He was overcharged $698.

3. Three-fourths of 292.

$\frac{3}{4}$ of 292

This means 292 times 3 and divide the result by 4.

$$\begin{array}{r} 292 \\ \times\ \ 3 \\ \hline 876 \end{array} \qquad 4\overline{)\,87{}^{3}6}\ \ \text{with quotient } 219$$

using short division

$\frac{3}{4}$ of 292 is 219.

4. $$\begin{array}{r} 1482 \\ -\,1457 \\ \hline 25 \end{array}$$

The doll named Dolly weighs 25 pounds.

5. $9^2 = 9 \times 9 = 81$

$5^2 = 5 \times 5 = 25$

$2^5 = 2 \times 2 \times 2 \times 2 \times 2 = 32$

$1^{387} = 1 \times \ldots \times 1 = 1$

Chapter Five
In the Saddle

Before Miss Ente showed up, Fred was wondering what to do for the next week. He created a function in which the domain was the set of all things that he might do and the codomain was {yes, maybe, no}.

To each member of the domain, he associated exactly one member of the codomain.

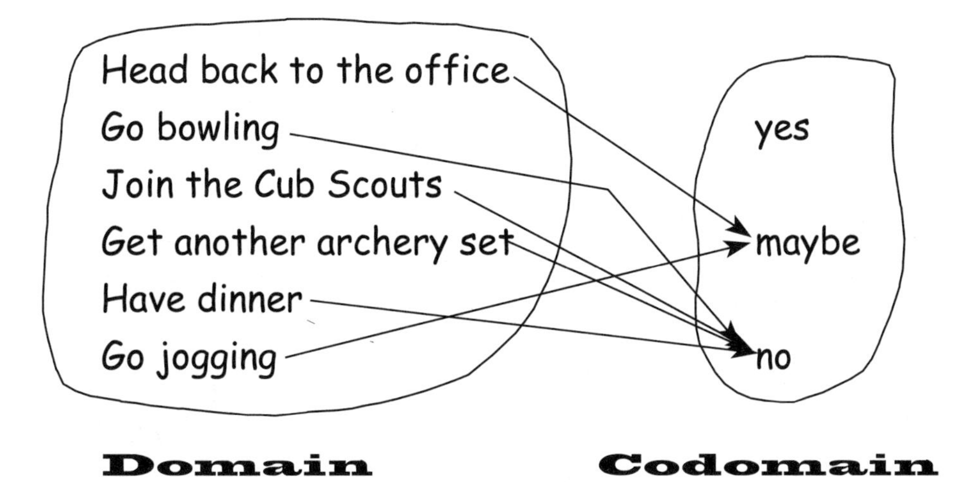

When Miss Ente showed up, she enlarged the domain of Fred's function. She added another alternative for possible things for Fred to do.

Fred's new function mapped *Head back to the office* and *Go jogging* to *maybe.*

It mapped *Go bowling, Join the Cub Scouts, Get another archery set,* and *Have dinner* to *no.*

And it now mapped *Go with Miss Ente to Camp Horsey-Ducky* to *yes.*

(Some day Fred might want to join the Cub Scouts, but he will need to be older.)

Fred was glad that he had already paid the camp fee, because he didn't have any more money with him.

"All your stuff is already at the camp," Miss Ente said. "It was all in the camp car that I towed to Horsey-Ducky."

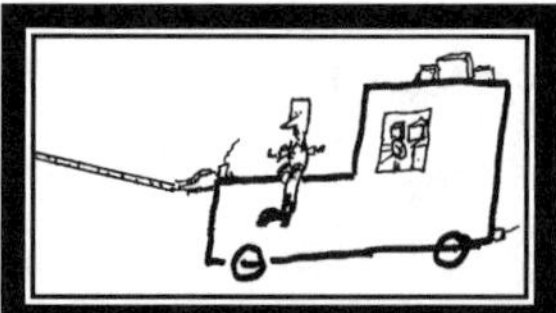

Photograph of the camp car being towed by Miss Ente (before Fred fell off)

"I fell off the hood," Fred admitted.

"That's okay. Everything is all right* now. Just hop in the saddle and I'll take you there."

Fred reaching →

Fred tried to reach up to the stirrup (the place where you put your foot), but he was too short.

He didn't have a ladder.

If he had a rope, he could lasso the saddle horn and climb up the rope, but he didn't have a rope.

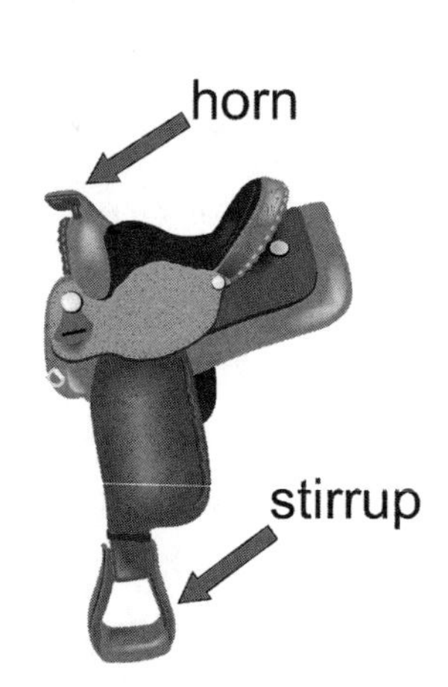

He looked around for someone to pick him up and put him in the saddle. There was no one.

Quiz Time!

Which of these inventions is super important in history?

A) peanut butter

B) tennis racquets

C) stirrups

D) Rag-A-Fluffy dolls

---

* *Alright* is not a word.

If you guessed Rag-A-Fluffy dolls, you get a grade of N.C. (= Not Close).

The correct answer is the stirrup. It was invented in about 500 A.D., and it changed the world.

Imagine a knight in heavy armor on a saddle without stirrups. One little push from the side and he would fall off his horse.

Note that his feet are not in stirrups.

With stirrups he could keep his balance while riding at full speed. A small number of knights could now be the equal of zillions of soldiers on the ground.

Stirrups "changed history, probably doing more to alter poltical institutions than any politician who ever lived. . . . The stirrup made feudalism not only possible but probably inevitable."*

But Fred was too short to reach the stirrup. He looked at the edge of the Great Lawn and found the perfect solution. He climbed up one of the trees, Miss Ente came over, and Fred jumped into the saddle.

Fred's feet did not reach the stirrups. He didn't want to fall off the way he had when he rode on the hood of the camp car. There were no seatbelts, so he held onto the saddle horn with both hands.

---

* That's what I learned on page 69 of *The Great Reckoning*, a book that I have read three times.

For 600 years feudalism (FEW-dal-ism) was the way that much of medieval Europe was structured. It's a big topic. In a single sentence: Feudalism was a society in which some people got to farm land in exchange for their service or labor.

"Are you ready?" Miss Ente asked.

Fred nodded but Miss Ente couldn't see him nod. Fred also wrapped his legs around the saddle horn, and Miss Ente galloped off toward Camp Horsey-Ducky.

Fred didn't waste the hour it took for them to get to Camp Horsey-Ducky. He thought of rules that might be functions. The first thing he thought of was the rule of ownership. Miss Ente owned Camp Horsey-Ducky and she owned her doll named Dolly. Fred owned Kingie.

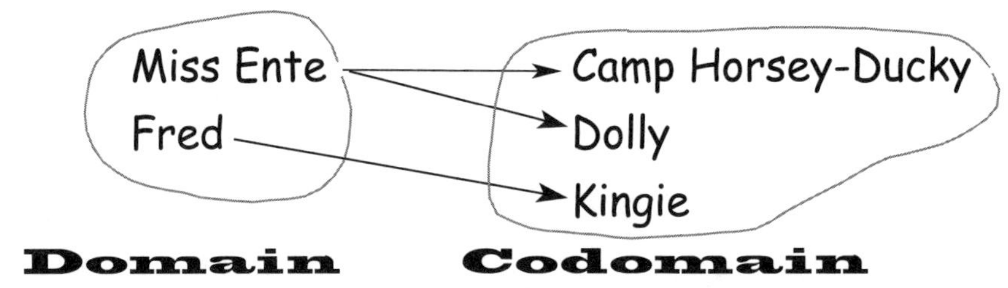

Is this a function? The definition of a function is that each element in the domain must be assigned to exactly one element in the codomain. Miss Ente is assigned to two different elements in the codomain. It is not a function.

Then Fred thought of the rule: *Assign to my four favorite students the state in which they currently live.*

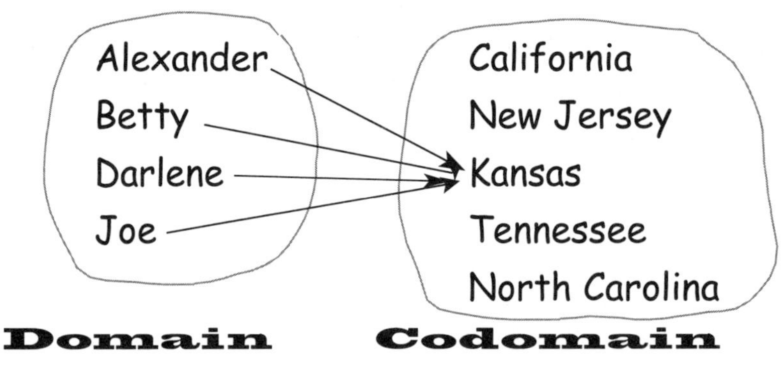

Is this rule a function?

Does each member of the domain have exactly one image in the codomain? Yes. It's a function.

Your Turn to Play

These all mean the same thing:

☞ Joe is a member of the set of all students at KITTENS University.

☞ Joe is an element of the set of all students at KITTENS University.

☞ Joe $\in$ the set of all students at KITTENS University.

These all mean the same thing:

☞ Fred is not a member of the set of all students at KITTENS University

☞ Fred is not an element of the set of all students at KITTENS University.

☞ Fred $\notin$ the set of all students at KITTENS University.

1. = is the symbol for "is equal to." Make a guess what the symbol for "is not equal to" is.

2. Fill in each blank with either $\in$ or $\notin$.

Joe ___ the set of teachers at KITTENS.

Fred ___ the set of teachers at KITTENS.

Fred ___ the set of all people who like to eat five pounds of jelly beans for lunch.

Joe ___ the set of all humans who are shorter than eight feet tall.

.......COMPLETE SOLUTIONS.......

1. If ∈ means "is an element of," and ∉ means "is not an element of," then a really good guess would be that ≠ means "is not equal to." And that guess would be right.

2.

Joe _∉_ the set of teachers at KITTENS. Joe is a student; he is not a teacher.

Fred _∈_ the set of teachers at KITTENS. Fred is a teacher at KITTENS.

Fred _∉_ the set of all people who like to eat five pounds of jelly beans for lunch. Fred would not even like two ounces of jelly beans. It would probably make him sick.

Joe _∈_ the set of all humans who are shorter than eight feet tall. Almost every person is a member of this set.

This is fun. Let me play a little bit.

C.C. Coalback ∉ the set of all honest people.

Methhunda ∉ the set of people who never lie.

Fred ∉ the set of all people who live in South Carolina.

Fred ∈ the set of all people who love mathematics.

Darlene ∉ the set of all ducks at the Great Lake on the KITTENS campus.

"$7 \times 8 = 56$" ∈ the set of all true statements.

"$0^{348} = 17$" ∉ the set of all true statements.

0^{348} equals $0 \times \ldots \times 0 = 0$

Chapter Six
Long Straight Road

Methhunda needed to get out of town as quickly as possible. If Coalback was released from jail (or escaped), she knew that he would be looking for revenge. Revenge is not something that good people do. Hurting other people is something that Coalback has done all his life.

She picked a bus stop at random and got off the bus. The stop was near a blood bank. It was a quarter after six in the evening.

"This is perfect," she said to herself. "Coalback would never get near a place where you can *give* to others. In fact, I'm also a little uncomfortable here."

The bus had been on the main highway. Perpendicular (per-pen-DICK-you-ler) to that highway is a long straight road that seems to lead to nowhere. She started walking down that road.

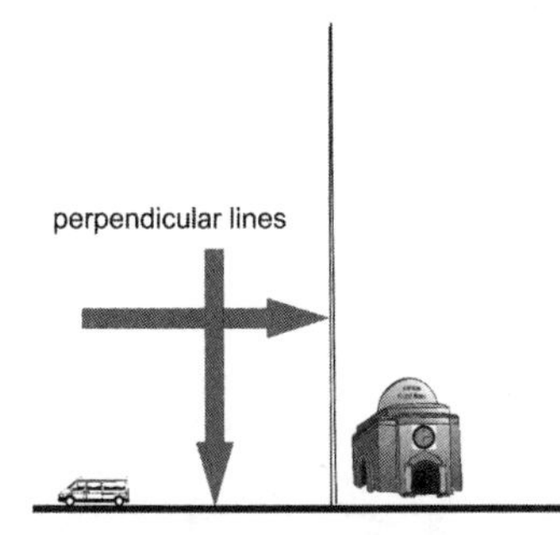

the long straight road

There is only one thing at the end of that long straight road: Camp Horsey-Ducky.

Mr. and Mrs. Gans were driving their son to Camp Horsey-Ducky. They had just turned off the main highway when their son exclaimed, "Look! There's a black garbage bag walking down the road!"

"That's not a garbage bag," his mother explained. "It's a woman who is wearing garbage sacks. That must be some new fad."

They stopped and Mr. Gans asked, "Would you like a ride?"

Methhunda said, "Yeah. Why not?" and climbed into the car.

Mrs. Gans asked her, "Are you also heading to Camp Horsey-Ducky?"

Methhunda mumbled, "Yeah. I guess so."

Obviously, this hitchhiker didn't want to talk very much. It became quiet in the car.

On the road they passed a little boy riding a large horse. The Ganses' son said, "That kid really doesn't know how to ride a horse. He's holding onto the the saddle horn with both hands and both legs. He must be very frightened."

Mr. Gans said, "He does appear a little anxious."

Methhunda said to herself, "That kid looks like the one I sold the $700 archery set to."

Very soon, the Ganses arrived at the entrance.

Methhunda said, "I hope we are not too late for dinner."

The son said, "It's two quarters after six."

"You mean its half past six," his mother said.

The son, whose name was Dumwalt, argued, "When we passed the blood bank, we all said that it was a quarter after six."

clock at the blood bank

"That's right, Dumwalt," Mrs. Gans said. "It was." She took out a piece of paper and drew a clock and cut it into quarters.*

Mrs. Gans's drawing of a clock cut into quarters

Dumwalt shouted, "Then I'm right! It must be two quarters past six right now."

right now

Dumwalt's mom explained to him, "Two-quarters is the same as two-fourths ($\frac{2}{4}$). Do you remember how to reduce fractions?"

Dumwalt shrugged his shoulders and said that he didn't.

"If you want to reduce $\frac{5}{15}$ you divide top and bottom by 5 and get $\frac{1}{3}$" she began.

---

* English is a lot harder than math.

one-half = $\frac{1}{2}$

one-third = $\frac{1}{3}$

but one-fourth = one quarter = $\frac{1}{4}$

one fifth = $\frac{1}{5}$

Why does one-fourth get two names? It's crazy.

"If you want to reduce $\frac{14}{21}$ you divide top and bottom by 7 and get $\frac{2}{3}$ and if you want to reduce $\frac{8}{27}$ you can't. There is no number that divides evenly into both 8 and 27."

By the time that Mrs. Gans had explained to her son that $\frac{2}{4}$ can be reduced to one-half ($\frac{1}{2}$) and that everyone says *half past six* instead of *two-quarters past six*, Miss Ente and her rider arrived.

"Where is he?" Mr. Gans asked.

"Dad, if you look carefully," Dumwalt said, "you can see that he is clutching the saddle horn."

"You are right, Dumma," Mr. Gans said. He carefully lifted Fred off of the saddle and put him on the ground.

small essay

Diminutives
(deh-MIN-you-tives)

Dumma is a diminutive form of Dumwalt. Kitty is a diminutive form of Katherine. Bobby is a diminutive form of Robert. Tommy is a diminutive form of Thomas.

Dumma

Right now, Dumma seems to be a much better name for him than Dumwalt. When he grows up and starts a large business called Dumwalt Gans Enterprises, then only his close friends will call him Dumma.

end of small essay

Your Turn to Play

1. Mrs. Gans drew a clock and cut it into fourths. Draw a circle and cut it into thirds.
2. Is this a function?

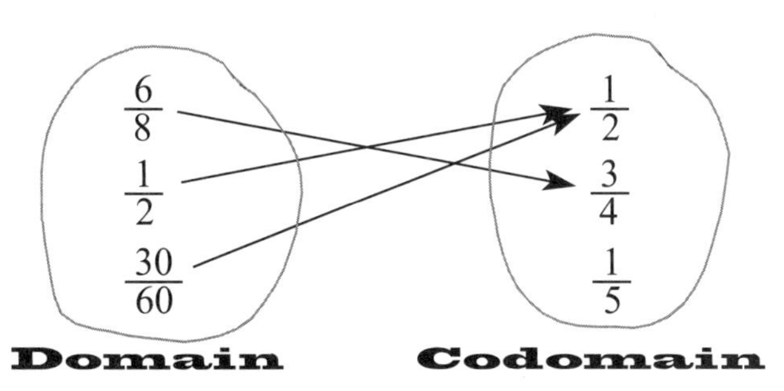

3. This is a map of Camp Horsey-Ducky. There are four rose gardens on the property. Miss Ente is going to have an 8-foot fence built around the edge of the property to keep the deer out. (Adult deer can jump over a 6-foot fence, and deer love to eat roses.)

Find the perimeter (the distance around) the property.

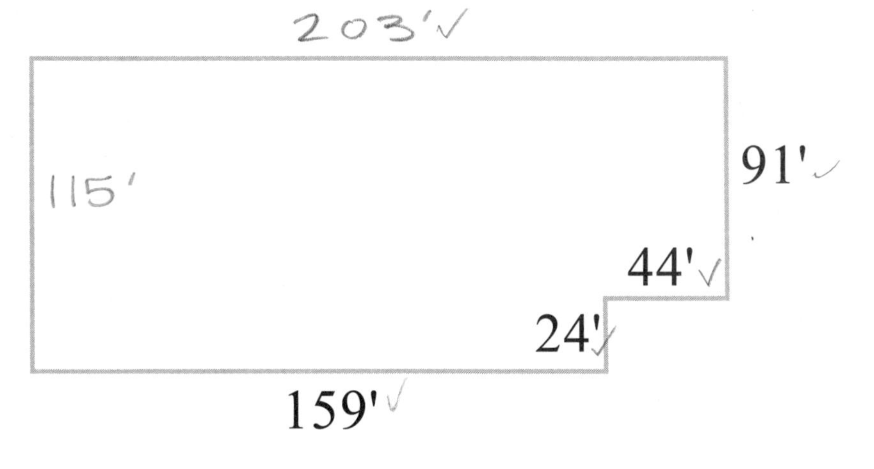

.......COMPLETE SOLUTIONS.......

1. This is a circle cut into thirds.

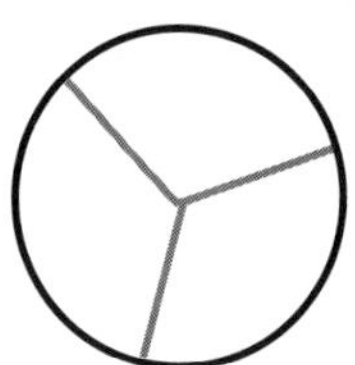

When six people share a pizza, it looks like this.

When eight people share a pizza, it looks like this.

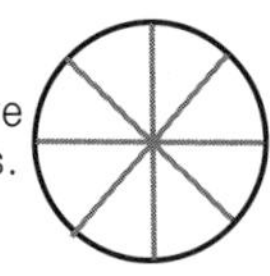

2. Does $\frac{6}{8}$ have exactly one image in the codomain? Yes.

Does $\frac{1}{2}$ have exactly one image in the codomain? Yes.

Does $\frac{30}{60}$ have exactly one image in the codomain? Yes.

Since each element of the domain has exactly one image in the codomain, it is a function.

(This function could have been described by the rule: *Reduce each fraction as far as possible.*)

3. The first thing we have to do is find the lengths that were not drawn in on the map.

The length marked by "?" is equal to 159 + 44 which is 203'.

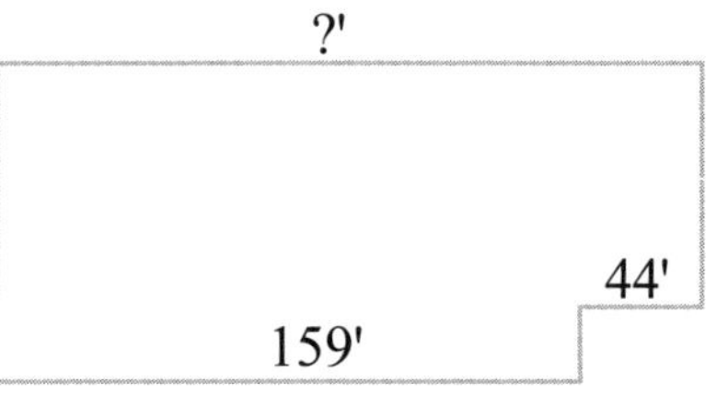

The length marked by "?" is equal to 91 + 24 which is 115'.

Now we know all the lengths. We add them.

$$\begin{array}{r} {}^{2\,2}203 \\ 91 \\ 44 \\ 24 \\ 159 \\ \underline{115} \\ 636 \text{ feet} \end{array}$$

Chapter Seven
At Camp

Fred ran to the camp car and unloaded his twelve boxes of camp equipment, his math books from the library, and his suitcase (which was a lunch box with a duck on it).

camp car

His twelve boxes included
an extra-small cowboy hat, neckerchief, rope for cows, gloves to avoid rope burns, silver spurs (with gold trim), harmonica, sundial (in case there were no clocks at the camp), mosquito spray, sun screen, campfire songbook, kite with string and extra tails, compass, a tent, bandages, ax, canteen, a sleeping bag, poison oak soap, lantern, pancake turner (for cooking breakfast in the great outdoors), six iron frying pans of various sizes, and a case of flares (to signal for help in case of an emergency).
He was already wearing his cowboy boots.

Meanwhile, Miss Ente was welcoming the other guests.

Mr. Gans introduced his son, Dumwalt, and said, “I hope that he will have a fun time at camp and learn a lot.”

“I’m sure he will,” Miss Ente replied.

Mr. and Mrs. Gans got in their car and left.

Methhunda just stood there. Miss Ente waited. As Methhunda looked around, her garbage bags rustled.

small essay

Silent Letters

Rustle is pronounced RUSS-el. The *t* is a silent letter. When you say the word *rustle*, you don't hear the *t*.

There are several words with silent *t* in them: listen, castle, soften, whistle and Christmas.

Is *t* the only letter that can be silent in English? No. In fact, most letters can occasionally be silent. This makes it extra hard to learn spelling.

a in spread, heaven, and cocoa

b in dumb, comb, bomb, and debt

c in brick, scene, muscle, and yacht

d in badge, fudge, handkerchief, and Wednesday

e in giraffe

f is pretty rare. The only one I could think of is halfpenny (HAY-penny), which is an old British coin.

g in sign, align, high, though

h in what, where, wheat, choir, and ghost

i in juice, and sluice

k in knife, knight, knee, knot, and knowledge

l in half, salmon, should, talk, and yolk

m in mnemonics (knee-MON-ics), which describes ways of improving your memory

n in autumn, column, and hymn

o in people, jeopardy, and leopard

p in raspberry, receipt, psalm, and psychology

r in February (but some people pronounce the *r*)

s in island and debris

t (I already did this one at the top of the previous page.)

u in guess, guide, build, and tongue

w in sword, two, answer, whole, wrist, and write

z in rendezvous

I couldn't think of any with *j*, *q*, *v*, *x*, or *y*.

end of small essay

Methhunda finally said, "Hey. I need a place to hang out for a while."

"I don't understand what you mean by hanging out. This is private property," Miss Ente said.

"Can't I just look around?"

"This is my home. You don't walk up to someone's home and ask to look around."

"No. I mean. . . . Like I think. . . ."

Miss Ente asked, "Are you interested in getting a job here at Camp Horsey-Ducky?"

Methhunda was never interested in working. She looked around and saw that there wasn't anything she could steal. She said, "See you later," and left.*

The men from the fence company arrived. They got out and were looking for the owner. All they could find was a horse.

One of them said to the other, "I just talked with the owner this morning. She was very nice and told us she'd meet us here this evening. I wonder where

* The only thing Methhunda was ever interested in was . . . herself.

she is. She said that she had brown hair and was a little over six feet tall."

Miss Ente said, "Hi."

The men fell silent. They looked at each other. They looked at the horse. They waited.

Miss Ente was used to this reaction. At least the men didn't faint like Fred did. She began slowly so that they could get used to hearing a horse talk. "Gentlemen, welcome to Camp Horsey-Ducky. I'm the owner that you spoke with on the phone. The eight-foot fence that you will be installing will be 636 feet long and will go around the perimeter of the camp."

She handed them a map of the property.
"It's almost the shape of a rectangle."

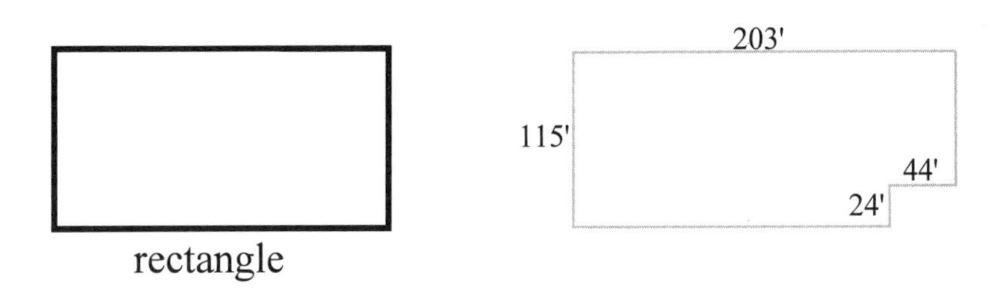

One of the men said, "It doesn't matter what the shape is. The 636 feet can be in the shape of a square, a triangle, or a rhombus (ROM-bus). We can fence any of those shapes."

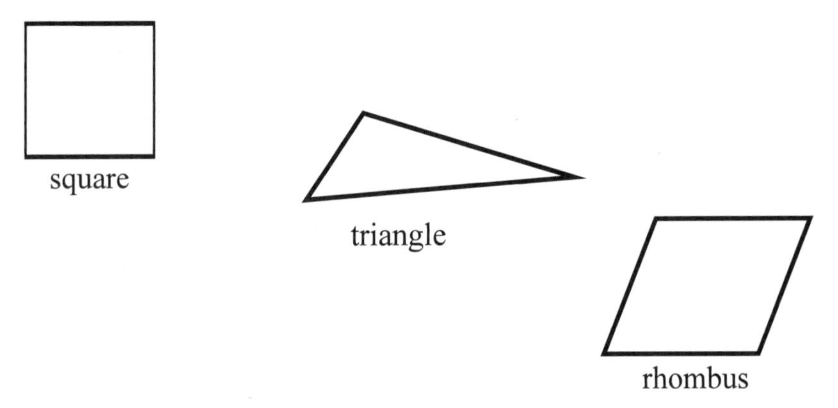

Your Turn to Play

1. The word *rhombus* has a silent letter. Which one is it?

2. Any figure with three (straight) sides is a triangle.
A rhombus is a figure with four equal sides.
Is every square a rhombus?

3. Is every rectangle a rhombus?

4. A rectangle is a four-sided figure with four right angles. Another way of saying that is: A rectangle is a four-sided figure in which any two sides that touch each other are perpendicular.

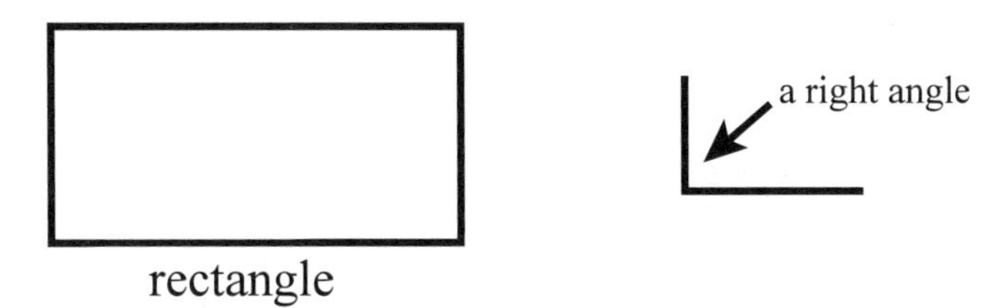

rectangle

Is every square a rectangle?

5. (Harder question) If something is both a rectangle and rhombus, must it be a square?

.......COMPLETE SOLUTIONS.......

1. The *h* in rhombus is silent.

2. *Is every square a rhombus* is the same as asking if every square has four equal sides. Yes. Every square is a rhombus.

3. *Is every rectangle a rhombus* is the same as asking if every rectangle has four equal sides. No. All four sides of a rectangle do not have to be equal. They can be, but they don't have to be.

4. *Is every square a rectangle* is the same as asking if every square has four right angles? Yes.

5. Suppose something is a rectangle. Then it must have four right angles.

Suppose something is a rhombus. Then it must have four equal sides.

If something is both a rectangle and a rhombus, then it must have four equal sides and four right angles. That forces it to be a square.

Can a four-sided figure have exactly three right angles?
Can a four-sided figure have exactly three equal sides?

The answer to one of these two questions is yes and the answer to the other is no.

Later in the book, I'll give the answers.

Chapter Eight

Playing with a Friend

Dumma went over to play with Fred. "What do you got in the boxes?" he asked Fred. Fred opened one of the boxes and showed him his six iron frying pans.

Dumma was amazed. "You must eat a lot to need six frying pans." He obviously didn't know Fred very well.

Dumma asked, "Did you bring a lot of food?" What he was hoping was that Fred had brought a lot of candy.

"No, I didn't. In fact, I didn't bring any food. I was expecting that we would live off the land. We could gather nuts, fruits, and berries."

Dumma was confused. "But you don't need an iron frying pan to cook nuts, fruits, or berries. That's silly."

Fred also wasn't sure why he brought those pans. The best that he could think of was, "Maybe if we find some hamburger out there in nature, we can cook it."

That seemed logical to four-year-old Dumma.

Dumma changed the subject. "Let's go and hunt tigers."

Fred was not used to playing with four-year-olds. For most of his life, he talked with college students and other grownups. Four-year-olds can switch topics quickly. One moment they are talking about frying pans, and a second later, they are talking about hunting tigers.

Fred looked through all his boxes. He didn't have anything with which to hunt tigers. He didn't realize that Dumma was talking about pretend-hunting, not real hunting.

Dumma put one of the frying pans on his head as a pretend-helmet. Fred did the same. Dumma's frying pan kept falling off. Fred's didn't. He had a flat head that was perfect for a frying-pan helmet.

They marched off together toward one of the rose gardens at Camp Horsey-Ducky.

Dumma said, "It's in the shape of a piece of pie." (Is the *e* in pie a silent *e*?)

Fred knew that in mathematics, that shape was called a sector.

small essay

Math Talk and Everyday Talk

At grandmother's house don't ask her for another **sector** of apple pie. Some grandmothers have never read any of the *Life of Fred* books. Instead, ask her for a piece of pie.

If she asks you how many pieces of candy you ate out of her candy dish, don't say that the **cardinality of that set** is equal to four. Just say you ate four pieces.

And if she wants an answer in writing, please do not write: cardinality { , , , } = 4. Just write, "I had four delicious pieces from the candy dish."

If you write a Valentine's note to her, do not write, "My love for you is > pizza." Just tell her that you like her more than pizza.

If she asks you if you have been good, don't tell her that your misdeeds are equal to the **empty set** { }.*

Naturally, if your grandmother has read *Life of Fred*, you can use all the math talk that you want. She will understand.

end of small essay

---

* Of course, if the cardinality of the set of your misdeeds is equal to one or two, don't lie to her. Just tell her that you have done one or two bad things.

Dumma asked Fred, "Do you see any tigers hiding in this rose garden?"

Fred didn't understand about playing with a friend. This was all make-believe. He was supposed to answer, "Yes. I see hundreds of them. They all look big and mean."

Instead, Fred said, "Of course not. There aren't any tigers running loose anywhere in Kansas."

This vexed* Dumma. He threw his frying-pan helmet on the ground and walked away. He didn't like it that Fred wasn't pretending. Dumma was acting like a four-year-old.

Fred picked up the frying pan and walked back to his boxes.

It was quiet. Fred looked around and wondered Where are all the other campers? He could see Dumma playing in the dirt near one of the water faucets. Dumma was making mud pancakes and pretending to eat them.

Miss Ente was just standing still, not doing anything.

Horses seem to do that a lot.

Fred walked up to Miss Ente and said, "Excuse me. If you have a moment, I have a couple of questions." Fred was taking advantage of the fact that Miss Ente could answer questions. Most horses pretend that they don't understand what you are saying.

"I've always wanted to know what horses are thinking about when they just stand there."

---

* To vex is to annoy, to bug, to bother, to nettle, to torment, to displease, to rile someone.

Miss Ente laughed. "I'm afraid that most of them are thinking about nothing. They are just standing there. Have you ever noticed how few horses ever achieve anything? They just eat and stand there.

"With a dog, you can throw a stick, and it will run and get it. With a horse, if you throw a stick, it will do nothing.

"With three-year-old kids, you can teach them the alphabet ABCDEFGHIJKLMNOPQRSTUVWXYZ*. Have you ever tried to teach a horse even one letter of the alphabet?

"What horse has ever written a song? Invented anything? Gone into business? Read a good book?"

Clearly, Miss Ente was ashamed of how few things most horses accomplish in life.

Intermission

When Miss Ente was young, her mother used to tell her that she should be different than most horses—and some people.

She would point to some people who spend their lives doing nothing except watching television, eating donuts, and going shopping.

"I want to be proud of you," she told Miss Ente. "Do something. Don't just be like other horses."

---

* In Greece the kids recite alpha, beta, gamma, delta. . . .
They write α, β, γ, δ. . . . Those are the letters of the Greek alphabet. Their alphabet is easier than ours. It has only 24 letters.

A lot of horses work for other people. Miss Ente had decided that she would be her own boss and start a business. She read a lot about how to run a business. She learned arithmetic, which you need in almost any business.

And she started Camp Horsey-Ducky.

Your Turn to Play

1. She was thinking about her income taxes when Fred interrupted her. She was adding up the expenses for her business.

advertising in THE KITTEN CABOODLE	$97.
business license	$86.
zoning permit	$69.

What were her total expenses?

2. What is the cardinality of {A, B, ε, ζ}?

In Greek, ε is epsilon and ζ is zeta.

3. Make a guess as to why we have presented the first six letters of the Greek alphabet—α (alpha), β (beta), γ (gamma), δ (delta), ε (epsilon), and ζ (zeta).

Choose one answer:

A) Because you will learn to speak Greek one day.

B) Because you will visit Greece some day.

C) Because the other 18 letters (24 – 6) are coming later.

D) Because in later mathematics we will use many of those letters.

4. Compute 87×94.

.......COMPLETE SOLUTIONS.......

1.
$$\begin{array}{r} 97 \\ 86 \\ +\ 69 \\ \hline 252 \end{array}$$
Her total expenses were \$252.

2. The cardinality of {A, B, ε, ζ} is 4.

The cardinality of a set is the number of elements in the set.

3. D) In later mathematics, we will use the Greek alphabet in many ways.

In trig, for example, we often use theta (θ) for angles.

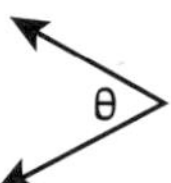

In the most important idea in calculus (the definition of "limit") we use both epsilon (ε) and delta (δ).

4.
$$\begin{array}{r} 94 \\ \times\ 87 \\ \hline 658 \\ 752\ \ \\ \hline 8178 \end{array}$$

or you could put the 87 on top →

$$\begin{array}{r} 87 \\ \times\ 94 \\ \hline 348 \\ 783\ \ \\ \hline 8178 \end{array}$$

You get the same answer in either case since $94 \times 87 = 87 \times 94$. Multiplication is commutative.

from the questions at the end of Chapter 7 . . .

Can a four-sided figure have exactly three right angles?

No, it can't. Once you have drawn three sides with right angles, the fourth angle will have to be a right angle.

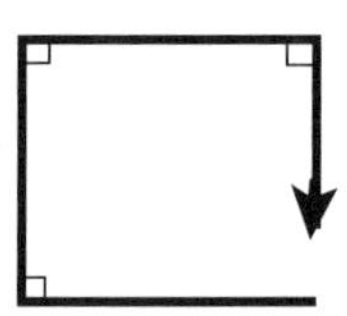

Can a four-sided figure have exactly three equal sides? Yes it can.

Chapter Nine
The Other Campers

Fred was amazed at Miss Ente's description of her early life. Fred had never read Prof. Eldwood's book entitled *All the Things Horses (and Some People) Have Not Achieved.*

Finally, Miss Ente said, "You said that you had a couple of questions. A couple means two. In the language of mathematics, the cardinality of the set of your questions is equal to two. In old-fashioned English, you told me you had a brace of questions."

Fred did the math in his head.

I had two questions.	2
I have asked one question	− 1
I must have one question I haven't asked	1

Fred was always astonished at how handy arithmetic can be. And he did not have to use a calculator.

"Oh yes." Fred remembered what his other question was. "Where are all the other campers?"

"There are five campers. You and Dumwalt are the only children. The other three arrived on the first day of camp."

Fred did a quick computation.

Total number of campers	5
Two of the campers are kids	− 2
Three must be adults	3

A million thoughts ran through Fred's mind: Are the adults out on an overnight hike? Are they around a campfire singing songs? Are they practicing at a rifle range? Are they in a class learning about rattlesnake bites? Are they at the camp cafeteria having dinner? Are they cooking dinner over a campfire? Are they pitching their tents and learning about the different knots that are needed around camp such as the bowline, the sheepshank, the taut-line hitch, and the square knot? Have they all gone out for some pizza? Are they learning how to play the guitar so that they can be like real cowboys?

the square knot

"The other three campers are already in the bunkhouse," Miss Ente said. "I'm sure that they will be glad to see you."

While Dumma continued to play in the mud, Fred raced to the bunkhouse to meet his new friends.

This was, perhaps, the first time in Fred's life that he had done arithmetic and had gotten the wrong answer! He had figured: 5 campers minus 2 kids equals 3 adults. This was not correct.

When Fred got to the bunkhouse, he discovered that he was the oldest camper at Camp Horsey-Ducky.

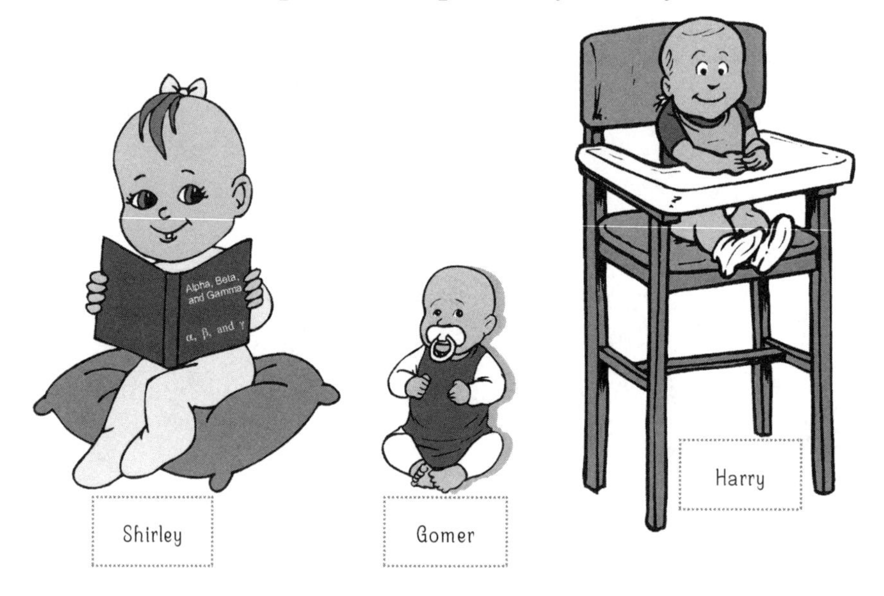

The correct arithmetic is: 5 campers minus 2 kids equals 3 babies.

Suddenly, Fred figured out why Miss Ente's business was called Camp Horsey-Ducky and not the Horse and Duck Club.

Miss Ente came to the door of the bunkhouse and looked inside. She asked, "How are we all doing?"

Shirley (the one with the book) said, "I read."

Gomer (the one with the pacifier) didn't say anything.

Harry (the one in the highchair) said, "I hungry."

Fred (the one with the square head) said, "I have a couple more questions. First, I have never heard of babies going off to camp for nine days."

"They are only here until ten o'clock each evening. Then their parents, Mr. and Mrs. Abend, come and pick them up. They both work until 9 p.m.* and then come, pick up their babies, and take them home. Only older kids like you and Dumma stay overnight."

Miss Ente knew what Fred's second question was going to be. She was polite and let him ask it.

He began his second question—and he knew already what the answer was going to be. "I know that you are a horse that achieves more than the average horse. You have both the ability and the drive to accomplish much."

Miss Ente interrupted Fred's second question. "That's exactly what my mother always told me. It's *ability* and *determination* that are the two keys to success. You need them

---

* Mr. and Mrs. Abend work at Edward's Afternoon Auto, which closes at 9 p.m. Mr. Abend helps Edward in the shop, and Mrs. Abend does the office work.

Edward, owner of Edward's Afternoon Auto

both. If you have only ability but no determination, you won't make it. I knew a horse that could speak four languages—English, German, French, and Horse—but he was lazy. He wasn't willing to work hard, and he got nowhere in life."

Fred thought of some of his students who would quit when they couldn't instantly answer a homework question. They weren't willing to generate even one drop of sweat. Those students usually never made it through college.

Miss Ente continued, "I knew another horse who had lots of drive. He would work for hours on the wrong things. He would work on projects for which he had no ability."

Fred giggled. He thought of his artistic ability compared with his doll, Kingie. Once, they each drew a picture of the same house. Can you guess which one was Fred's picture?

"But even with your ability and drive," Fred continued, "I do not see how you are able to take care of these three babies."

Miss Ente waited for Fred's second question. He still hadn't asked it. Clearly, he was afraid to ask it. He knew what the answer would be.

"What if, for example, Gomer's pacifier fell out of his mouth? How could you put it back in?" Fred asked.

Just then, that happened. And Gomer was getting ready to cry. Fred picked up the pacifier and reinserted it. That is something that can be done with hands but not horse hoofs.*

unhappy Gomer

"Who is going to take care of all these babies?" Fred finally asked.

Miss Ente looked at Fred, and Fred knew the answer.

Your Turn to Play

Fred knew mathematics. He could find out what two-sevenths ($\frac{2}{7}$) of 294 was equal to, but he didn't know anything about working with babies.

1. What word in, ". . . he didn't know anything about working with babies" has two silent letters?
2. What is two-sevenths of 294?
3. Fred looked at the clock on the wall. What time was it?
4. At 10 p.m. Mr. and Mrs. Abend would come to pick up their three babies. How long would Fred have to take care of them?
5. Dumma wandered into the bunkhouse. He was covered with mud. Fred mentally did the arithmetic (3 + 1 = 4). He had four kids to take care of. Dumma flopped into bed and fell asleep. Fred did the arithmetic (4 – 1 = 3). *Adding one* and *subtracting one* are called inverse operations. What is the inverse operation to *adding 800*?

* The plural of the word *hoof* is either *hoofs* or *hooves*. Both are correct.

.......COMPLETE SOLUTIONS.......

1. *Know* is pronounced NO. Both the *k* and the *w* are silent.
2. Two-sevenths of 294

   $\frac{2}{7}$ of 294

   294 times 2 and divide the result by 7

   $$\begin{array}{r} 294 \\ \times \quad 2 \\ \hline 588 \end{array} \qquad \begin{array}{r} 84 \\ 7\overline{)5\,8^{2}8} \end{array}$$ using short division

3.  either ten minutes to eight or 7:50
4. From 7:50 to 8:00 is ten minutes.

   From 8:00 to 10:00 is two hours.

   It is two hours and ten minutes from 7:50 to 10:00.
5. The inverse operation to *adding 800* is *subtracting 800*.

The inverse operation to subtracting 5 is adding 5.

The inverse operation to multiplying by 7 is dividing by 7. (If you multiply something by 7 and then divide your answer by 7, you will get back to where you started.)

The inverse operation to walking east (→→→) is walking west (←←←). You will get back to where you started.

There is no inverse operation to eating a pickle. You can't get back to the pickle before you chewed it and swallowed it.

Chapter Ten
Reading

Fred didn't know how to ice skate. He didn't know how to build an airplane. He didn't know how to mix and pour concrete in order to make a sidewalk.*

And he didn't know about caring for babies.

If he were back at his office, he would have pulled his copy of Prof. Eldwood's *Modern Baby Care*, 1846, off the shelf and read it. Here at Camp Horsey-Ducky, Fred had to guess what to do.

Shirley (the one with the book) said, "You read." Fred was relieved and thought At least I know something about reading. He looked at the book she was holding. It was an ABC book in Greek!

Fred looked around the bunkhouse. There weren't any other books. He was stuck with the one Shirley was holding.

When Fred first learned to read, his ABC book was in English:

A is for aerosol can.

B is for balalaika.

C is for colander.

It was a very hard book. Fred didn't know what an aerosol can was. His mother always just called it bug spray.

---

* Some people confuse concrete with cement. It is not quite correct to say cement sidewalks. They are concrete.

Cement cracks pretty easily. If you mix cement with sand and gravel, you get concrete, which is much harder to crack.

He had never heard of a balalaika (bala-LIE-ka), which is a Russian stringed instrument with a body in the shape of a triangle. Later, when Fred taught about triangles in his geometry class, he always mentioned balalaikas—and none of his students knew what he was talking about.

He had never seen a colander. He couldn't figure out what a metal bowl with holes in it was good for.

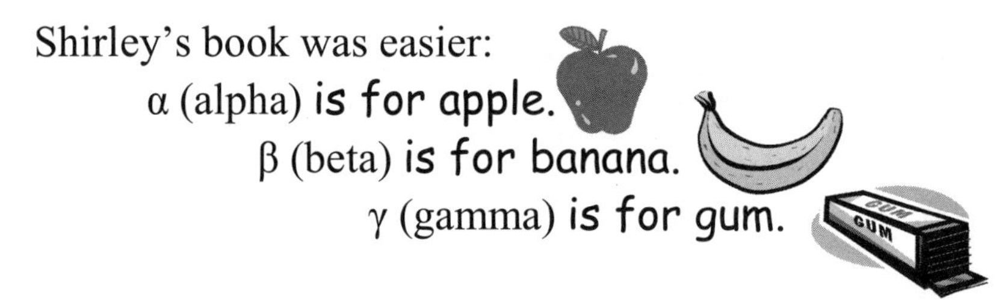

Shirley's book was easier:

α (alpha) is for apple.

β (beta) is for banana.

γ (gamma) is for gum.

She couldn't read the words yet. She just pointed to the pictures. Her favorite was the which she pronounced as banananananananananananananananananananana.

Gomer giggled when Shirley said banananananananana-nanananananananananananana. His pacifier popped out of his mouth. He hoped some day to be able to say that word. Today, he could only drool. Fred carefully wiped Gomer's mouth and reinserted the pacifier. Fred thought that pacifiers were just drool-plugs designed to keep babies from slobbering. (He was wrong.)

The big challenge would come from Harry the Hungry. He was already in his highchair. His hands were folded. He was waiting to eat.

Most of what Fred knew about food came from the nine vending machines in the hallway outside of his office. Fred would put in the coins, get the food, take it to his office, and put it in his desk "for later."

Fred looked around the bunkhouse . . . no vending machines! He panicked. There were no restaurants around. Fred's thoughts ran wild. What shall I do? We are all going to starve to death. We are going to have to eat the mud pancakes that Dumma made.

Harry knew more than Fred did. He pointed to a special room in the bunkhouse. It was marked **Bunkhouse Kitchen**.

The only kitchen that Fred could remember ever being in was the big kitchen at the university president's mansion. That fancy kitchen had six cooks in it.

Fred looked in the bunkhouse kitchen and counted the number of cooks. There were zero cooks. He told Harry, "The cardinality of the set of cooks now in the bunkhouse kitchen is equal to zero."

Harry said, "No cooks."

Fred asked, "If there aren't any professional cooks in the kitchen, how can you get fed?"

Harry said, "You cook."

small essay

Prolix versus Laconic

Harry was laconic (leh-CON-ik). He used very few words.

Professor Clef, whom we met in *Life of Fred: Kidneys*, was prolix (pro-LICKS). He was very wordy. He liked to talk and talk.

Imagine Professor Clef and Harry having a conversation. First, Clef asks a question: "I have traveled to almost all the countries of the world. Some of them are large and some are small. Some have cold climates. Some have lots of government laws, lots of taxes, and lots of regulations, and some are much more free. Next month, I was thinking of heading to see

Freedonia. The motto of Freedonia is **LIBERTY!** which means a special kind of freedom. Liberty is freedom from excessive government control. I'll be heading on a cruise ship, and I've been looking for someone to go with me on that trip. The ship is 534 feet long. There is a large library onboard, and each evening there are lectures on ancient Greece. Would you like to join me?

Harry answered, "Sure."

Prolix Professor Clef.
Laconic Harry.

end of small essay

"Me cook?" Fred replied laconically.

"Yep," Harry said.

"OK," Fred said. When he said this, he thought he had just won the Laconic Man of the Year Award.

Then Harry just pointed toward the kitchen, saying no words at all. Fred had been beaten.*

---

* It's fun to play the Laconic Game every once in a while, but it would be horrible to live with people who always spoke laconically.

You wake up and say, "It's a beautiful morning, isn't it?" and they say, "Um."

You tell them that yesterday you won the all-school math competition, and they say, "So?"

You ask them if they have seen your red sweater, and they just look at you and say nothing.

On the other hand, it might drive you nuts if you lived with a wordy-birdie who never stopped talking.

If you want to be liked, be neither laconic nor prolix.

Your Turn to Play

The cardinality of a set is the number of members in that set. The cardinality of {#, ♲} is 2.

1. This is the empty set: { }. It is the set that contains no members. What is the cardinality of { }?
2. The cardinal numbers are the numbers used to count the members of sets. What is the smallest cardinal number?
3. Suppose some crazy duck tells you . . .

How could you show that he was wrong?

4. The natural numbers are 1, 2, 3, 4, 5, 6, 7, 8. . . .
They go on forever. There is no largest natural number. Name a set whose cardinality is *not* a natural number.

5. What is the inverse operation to *putting on your hat*?

.......COMPLETE SOLUTIONS.......

1. The cardinality of { } is the number of members in that set. It has no members. The cardinality of the empty set is zero.

2. The smallest cardinal number is zero. It would be impossible to name a set with fewer members than { }.

3. All you would have to do is show that duck a set with seven or more members.

For example, {A, B, #, ☎, ✐, ✂, ❁}.

4. The smallest natural number is 1. You cannot find a natural number to describe the cardinality of { }.

Also you cannot find a natural number to describe the cardinality of all the natural numbers {1, 2, 3, 4, 5 . . .}.

5. The inverse operation to *putting on your hat* is *taking off your hat.*

And the inverse operation to *taking off your hat* is *putting on your hat.*

Chapter Eleven
Into the Kitchen

Fred gathered up all his courage and actually walked into the bunkhouse kitchen. Many things in life are scary the first time they are done:

the first time you meet a crazy duck,

the first time you ride a bicycle,

the first time you dive into a swimming pool.

This was Fred's first time attempting to cook in a kitchen. He couldn't even tell a colander from a carrot.

(KUL-en-der)

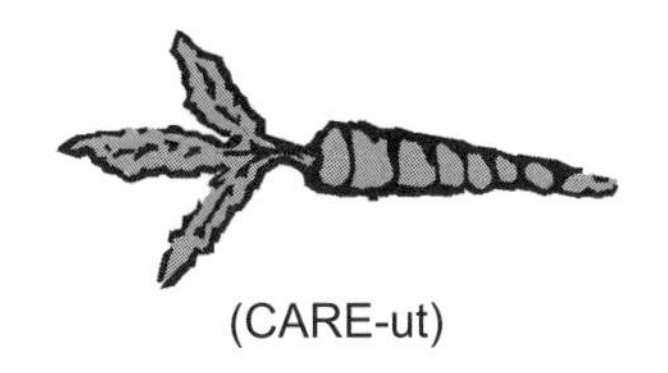

(CARE-ut)

He carefully opened one of the drawers. He didn't want to break anything. Inside were three white plastic cups. *How cute!* he thought to himself. *Those must be drinking cups for the three bears—one for papa bear, one for mama bear, and one for baby bear.** He took them out of the drawer and put them on the table.

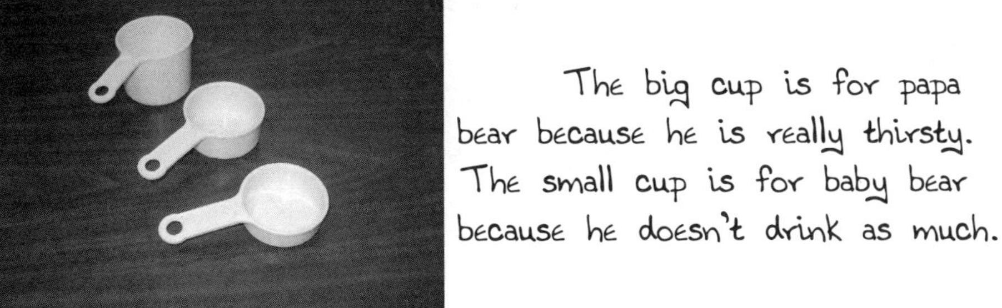

The big cup is for papa bear because he is really thirsty. The small cup is for baby bear because he doesn't drink as much.

* Fred was thinking of the story of Goldilocks and the Three Bears.

When Fred looked at the papa bear cup he thought *How silly! It says "one cup." Of course it is one cup. It is not two cups.*

Fred didn't realize that he was playing with measuring cups, not drinking cups. The measuring cups are used to measure volume.

In almost every country in the world, except the United States, the metric system is used. It is much simpler than the Imperial system as used in the United States.

U.S. system for volume measurement	Metric system for volume measurement
3 teaspoons = 1 tablespoon	1000 milliliters = 1 liter
2 tablespoons = 1 fluid ounce	
4 fluid ounces = 1 gill (also known as a half cup)	
8 fluid ounces = 1 cup	
2 cups = 1 pint	
2 pints = 1 quart	
4 quarts = 1 gallon	

One quart is almost a liter.* If you take a quart and add about three and a half tablespoons, you get a liter.

---

* When we get to *Life of Fred: Decimals and Percents*, we might say that a quart is approximately 94.63% of a liter, but we would never mention that in this book.

The mama bear cup was marked $\frac{1}{2}$ cup.

At this point Fred realized that it didn't mean taking a cup and cutting it in half.

One half of a pizza means

but $\frac{1}{2}$ cup does not mean

That would be silly.

Fred figured out that measuring cups are used for measuring volume. He filled the $\frac{1}{2}$ cup with water and poured it into the 1 cup. He did it again and that completely filled up the 1 cup.

$$\frac{1}{2} + \frac{1}{2} = 1$$

He emptied the cups and dried them off. Then he tried two half-cups of rice. That filled up the 1 cup.

He tried it with peanut butter . . . and it made a giant mess. The peanut butter stuck to the cup. It wouldn't pour. He used a knife to scrape it out. Then part of the peanut butter stuck to the knife. He used a second knife to scrape the peanut butter off the first knife. Some peanut butter got on his shirt.

$\frac{1}{2}$ cup peanut butter + $\frac{1}{2}$ cup peanut butter = a mess

Fred could hear Harry. If Fred had been six blocks away, he could have heard Harry.

He looked in the kitchen cupboards. There was no food there, just dishes and glasses.

He looked in the pantry. There was a sack of flour. He didn't know what to do with that. There were some jars of baby food. He tried to open one but couldn't. He decided *not* to use a hammer to get one of them open.

He looked in the refrigerator and found a carton of eggs. He took one of them and looked at it. He wasn't sure what you do with an egg. Maybe Harry can figure it out. He brought it out to Harry and put it on his tray.

Harry rolled it around on his tray.

As Fred headed back into the kitchen, he heard two things: splat! (as the egg hit the floor) and "I hungry!"

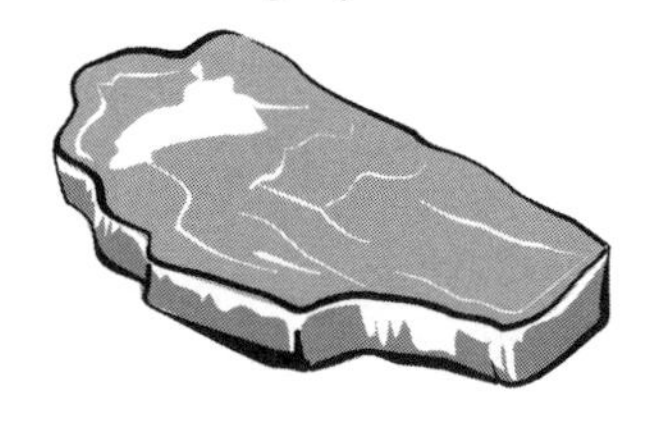

The only other thing in the refrigerator was a big red, juicy steak. Fred took it out of its plastic wrapping and put it on a plate. The steak was cold and red.

Fred brought it out to Harry. Shirley said, "Mommy always cooks it" and then she went back to reading.

Fred said, "Oops" and took it back to the kitchen. Somehow he knew that he had to figure out how to change the steak from cold and red to warm and brown.

On a shelf in the kitchen were a stack of cookbooks:

Prof. Eldwood's *Guide to Perfect Cream Puffs*
Prof. Eldwood's *Making Taffy*
Prof. Eldwood's *Milkshakes for Mathematicians*
Prof. Eldwood's *Deep-fried Apple Pie*
Prof. Eldwood's *Toothsome** *Steaks*

Titles of books are always *italicized.*

If you are handwriting, you underline them.

Your Turn to Play

1. Fred found that two half cups were equal to one cup.

$$\frac{1}{2} + \frac{1}{2} = 1$$

The baby bear measuring cup was marked "$\frac{1}{3}$ cup."

Make a guess: How many one-third cups would equal one cup.

2. $\frac{1}{4} + \frac{1}{4} + \frac{1}{4} + \frac{1}{4} = ?$

3. Find 70% of 1190.

We did this kind of problem in the previous book. Here is an example of how it is done:

Find 30% of 270.

$\frac{30}{100}$ of 270 — changing a percent into a fraction

$\frac{3}{10}$ of 270 — reducing the fraction by dividing top and bottom by the same number, in this case, by 10.

270 multiplied by 3 and divided by 10

$$\begin{array}{r} 270 \\ \times\ \ 3 \\ \hline 810 \end{array} \qquad 810 \div 10 = 81$$

30% of 270 is 81.

---

* Toothsome doesn't mean having some teeth. It means delicious or appetizing.

.......COMPLETE SOLUTIONS.......

1. $\frac{1}{3} + \frac{1}{3} + \frac{1}{3} = 1$

2. $\frac{1}{4} + \frac{1}{4} + \frac{1}{4} + \frac{1}{4} = 1$

3. 70% of 1190

$\frac{70}{100}$ of 1190

$\frac{7}{10}$ of 1190 reducing the fraction by dividing top and bottom by 10

1190 multiplied by 7 and divided by 10

$$\begin{array}{r} 1190 \\ \times \quad 7 \\ \hline 8330 \end{array}$$

$$\begin{array}{r} 833 \\ 10\overline{)8330} \\ \underline{80} \\ 33 \\ \underline{30} \\ 30 \\ \underline{30} \\ 0 \end{array}$$

70% of 1190 is 833.

You may have noticed a pattern for dividing by 10.

$40 \div 10 = 4$
$780 \div 10 = 78$
$39{,}850 \div 10 = 3{,}985$
$1{,}600 \div 10 = 160$
$10{,}000 \div 10 = 1{,}000$
$8330 \div 10 = 833$

÷ means divided by

Chapter Twelve
Harry's Needs

Fred pulled Prof. Eldwood's *Toothsome Steaks* off the shelf. He was ready to learn how to turn a cold and red steak into a brown and warm one. And he was hoping that it wasn't too complicated.

The first chapter listed all the kinds of steak.* That didn't matter to Fred since he had already discarded the plastic wrapper that stated what kind of steak it was.

The second chapter listed some steak sauce recipes to serve with steaks: Garlic Butter, page 832; Mushroom and Grape Juice, page 395; Béarnaise Sauce, page 503; Bordelaise Sauce, page 919; and Horseradish Dressing, page 492. Fred decided that he was going to serve the steak without any special sauce.

The third chapter described broiling steaks. The first words were, "Preheat the broiling compartment." Fred had no idea what a *broiling compartment* was. He knew what a *broiling hot apartment* was. That was an apartment in the summertime without air conditioning. He skipped the third chapter.

---

* For example: Top Blade Steak (also known as Book Steak or Butler Steak), Shoulder Steak (also known as English Steak), Chuck Arm Steak (also known as Round Bone Steak), Ribeye Steak (also known as Delmonico Steak), T-Bone Steak (also known as Porterhouse), Filet Mignon, Top Loin Steak (also known as New York Strip Steak or Ambassador Steak), Sirloin Steak, Tri-Tip Steak, Round Tip Steak, Round Steak (also known as Top Round London Broil), Skirt Steak (also known as Philadelphia Steak), Hanger Steak, and Flank Steak (also known as Jiffy Steak or London Broil).

The fourth chapter offered some hope: Cooking Steak in a Pan. *Yes!* Fred thought. *This I can understand.* He ran out to his box of iron frying pans and selected one of them. He picked one that was a little larger than the steak.

As he ran back to the kitchen, Shirley said, "Harry hungry." Harry was getting ready to cry. Fred patted Harry on the head with the hand that wasn't carrying the frying pan and said, "I'm cooking you dinner right now."

Back in the kitchen, he started reading the fourth chapter: Heat pan with a lively heat.

What does "lively heat" mean? he wondered. He rushed to the dictionary in the main room in the bunkhouse and read that *lively* means full of energy or exciting or sparkling or vivid.

Harry was now crying softly.

Back in the kitchen, Fred put the pan on the stove and turned on the burner. The dial read warm—medium—hot. The *medium* didn't seem lively enough and the *hot* seemed more than lively, so he set the dial halfway between medium and hot.

Then he read: Stick steak into pan and sear each side for five minutes.

Fred was confused. *What does "sear" mean?* Fred ran back to the dictionary and looked up the word *sear*. He found out that sear was exactly what he was doing to the meat right now.

Harry's crying could now be described as lively. Fred ran back to Harry and gave him another pat on the head. That didn't help. He offered to read a book to him. He offered to do a little dance to entertain him.

Wait a minute! I, your reader, have to object. Why is Fred doing all the stuff when it is obvious that what poor little Harry wants is food? Why doesn't Fred stay in the kitchen?

I, your author, do not want to interrupt this story right now. Too much is happening.

I, your reader, am in charge. I'm the one holding the book in my hands. Answer me. Why is Fred messing around with patting Harry on the head instead of feeding him?*

You are asking a BIG question. This is usually dealt with in the study of psychology (the science of the mind and mental states) at the college level.

Now! (an imperative sentece)

Okay. I'm going to have to simplify a lot.

I'm going to used this type face so that you can skip all this psychology if you like and get back to the story.

There are three main approaches to psychology:

1. Mr. Freud's approach, which dealt with all the things that can go wrong—the immature, the stunted, and the unhealthy. Personality disorders are the favorite area of study in first-year psychology classes.
2. Mr. Skinner's approach, which treats people as if they were rats or robots. You change behavior by offering rewards and punishments.
3. Abraham Maslow's approach, which changed everything. Instead of studying the nut cases or the rat-brained, he looked at the very healthiest humans. He didn't concentrate on the alcoholics, the violent, or the criminal, but on those who have the most joy in life.

Okay. But what does this have to do with Fred patting Harry on the head instead of feeding him?

---

* This is such a perfect example of three of the four sentence patterns.

I am the one holding the book in my hands. A declarative sentence = a sentence that states something.

Answer me. An imperative sentence = a sentence that is a command.

Why is Fred messing around? An interrogative sentence = a question.

In a few moments, Fred may say, "Oh!" which is an exclamation, the fourth sentence pattern.

I'm getting to that. I told you that this is a BIG question.

Maslow noticed that there were five steps that we take in becoming our best. They are done in this order. ←important!

Step 1: Body needs. Food, sleep, water, air. This comes first.
Step 2: Safety needs. Nobody is going to beat you up or scream at you. You have enough money.
Step 3: Love needs. You get hugs. You have friends.
Step 4: Esteem needs. People admire you because of what you have done. Recognition and awards.
Step 5: The place where all those other needs fade in importance. Maslow called it self-actualization. It is achieved by very few people. They do not consider themselves to be the center of the universe. Their goal is not to get (food, money, hugs, awards) but to give.

So the short answer is that Fred was ignoring Maslow's five steps. Harry had a need at Step 1. He wanted food. Fred was offering Step 3. He offered a pat on the head.

If you were being attacked by an angry pangolin (Step 2 safety need), would you be in the mood to receive a trophy for your video skills (Step 4 esteem need)?

a pangolin

Pangolin is in your dictionary.

Your Turn to Play

1. Pangolins (PANG-geh-lynns) really exist. They are mammals that have horny scales. They live in Africa and Asia. If you ask 600 people if they know what a pangolin is, maybe only two percent of them have ever heard of a pangolin.

What is 2% of 600?

2. If you ask 600 people what a mammal is, ninety percent of them can tell you that mammals are animals that:

✓ are warm-blooded. (Fish are not-warm-blooded. If you put a thermometer into the mouth of a fish, its temperature will be the same as the water it is swimming in.)

✓ are air-breathing.

✓ are vertebrates. (They have a spine and a skull.)

✓ have young that receive milk from their mothers.

✓ have hair.

What is 90% of 600?

3. Fill in each blank with either $\in$ or $\notin$.

Fred ___ the set of all mammals.

A piece of spaghetti ___ the set of all mammals.

A statue of the president of KITTENS University ___ the set of all mammals.

Your teeth ___ the set of all mammals.

.......COMPLETE SOLUTIONS.......

1. 2% of 600

$\frac{2}{100}$ of 600

$\frac{1}{50}$ of 600 reducing the fraction by dividing top and bottom by 2

600 multiplied by 1 and divided by 50

$600 \times 1 = 600$

$$\begin{array}{r} 12 \\ 50\overline{)600} \\ \underline{50} \\ 100 \\ \underline{100} \\ 0 \end{array}$$

2% of 600 is 12.

2. 90% of 600

$\frac{90}{100}$ of 600

$\frac{9}{10}$ of 600 reducing the fraction by dividing top and bottom by 10

600 multiplied by 9 and divided by 10

$$\begin{array}{r} 600 \\ \times \quad 9 \\ \hline 5400 \end{array}$$

5400 divided by 10 is 540.
We showed how this is done at the end of the previous Your Turn to Play.

90% of 600 is 540.

3. Fred _$\in$_ the set of all mammals. (All humans are mammals.)

A piece of spaghetti _$\notin$_ the set of all mammals.

A statue of the president of KITTENS University _$\notin$_ the set of all mammals. (Statues are made out of stone. They are not warm-blooded or air-breathing.)

Your teeth _$\notin$_ the set of all mammals. (Your teeth are not vertebrates. Your teeth do not have hair. They are part of a mammal, but they are not a mammal.)

Chapter Thirteen
Smoke

The steak was in the pan. The pan was on the stove. The burner was turned to medium-hot. Fred was in the other room attempting to stop Harry from crying. I, your author, was answering your BIG question regarding the five different kinds of needs that all humans have. Somehow, we were talking about pangolins. You did a Your Turn to Play.

Smoke filled the room.

Shirley started coughing.

The smoke alarm screamed.

Gomer put a pacifier in each ear because of the noise.

For once, Fred did not panic. He did not faint. He picked up Gomer and took him outside. Then he picked up Harry and took him outside. He asked Shirley go outside and join her brothers. She took her book along. He woke up Dumma and asked him to go outside.

The safety of the kids was paramount. (*Paramount* = overriding, most important.) Fred didn't know if the bunkhouse was on fire. Moving the kids to a place where they would not be in danger was the first thing to do.

Fred went into the bunkhouse kitchen and turned off the stove. The steak was no longer cold and red. It was not warm and brown. It was hot and black.

And tiny. Fred had turned a 4-pound, 2-ounce steak into 3 ounces of garbage.

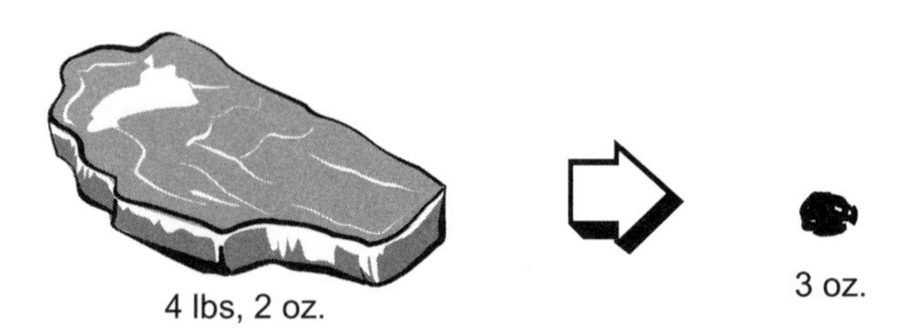

How much weight did the steak lose? The General Rule: if you don't know whether to add, subtract, multiply, or divide, is to restate the problem with simpler numbers. Suppose the steak went from 8 ounces to 3 ounces. It lost 5 ounces. We subtracted.

So we need to subtract 3 ounces from 4 pounds, 2 ounces.

	4 lbs.	2 oz.
−		3 oz.

The difficulty is that it is tough to take 3 away from 2. We have to borrow from the 4 lbs. Here is how it is done:

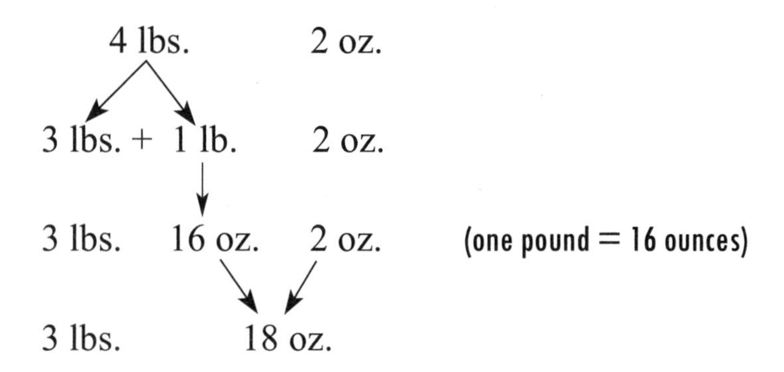

When people do

	4 lbs.	2 oz.
–		3 oz.

it often looks like this:

	3	16 + 2
	~~4~~ lbs.	~~2~~ oz.
–		3 oz.

and then the subtraction is easy.

	3 lbs.	18 oz.
–		3 oz.
	3 lbs.	15 oz.

The steak lost 3 pounds, 15 ounces.

The steak had been experiencing a lively heat for 20 minutes and 10 seconds. For the first 7 minutes and 25 seconds, it had *not* been burning. How long had it been burning?

	20 minutes	10 seconds
–	7 minutes	25 seconds

Since we can't subtract 25 from 10, we have to borrow a minute from the 20 minutes.

	19	60 + 10
	2~~0~~ minutes	1~~0~~ seconds
–	7 minutes	25 seconds
	12 minutes	45 seconds

When we borrowed one minute, that was equal to 60 seconds.

Fred opened the windows to let the smoke out. Soon the smoke alarm became silent.

Shirley took Gomer's two pacifiers out of his ears. She was a good sister.

Shirley $\in$ the set of all good sisters.

She went back to reading her book.

Fred went outside and picked up Gomer and brought him inside. He wondered where Gomer had gotten his second pacifier.

Shirley picked up Harry, took him to his bed, and tucked him in. Harry had cried himself to sleep.

Dumma was playing in the mud again. In a minute or so, he realized that he was tired. He came in and flopped into bed for the second time this evening.

Fred now had some cleanup to do. The smoke damage was everywhere. Originally, the bunkhouse looked like this.

Now, this was the mess that Fred faced. The walls would have to be washed. The carpets would have to be cleaned. The plant would have to be replaced.

Your Turn to Play

Fred started with the wall on the far side where the plant was. He dragged the plant outside and threw it into the garbage.

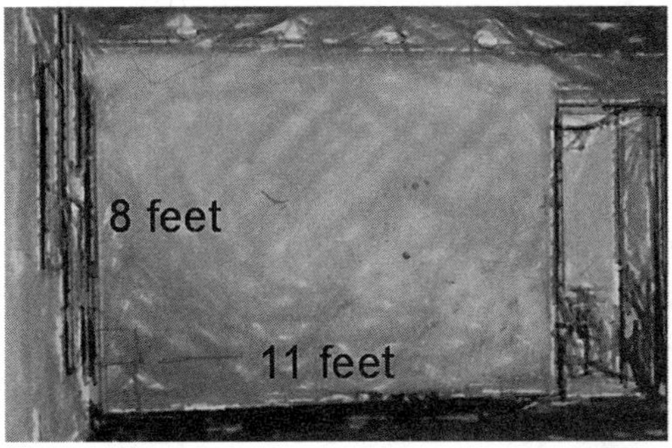

He looked at that smoky wall. It was a rectangle that was 8 feet by 11 feet.

1. What was the area of that wall?

2. What was its perimeter?

3. Fred was going to have to wash the wall to get rid of the smoke. Which was more important to him: the area or the perimeter of the wall?

4. He filled a bucket with hot soapy water. It held 4 gallons of water. It was too heavy for Fred to carry. He poured out 2 gallons, 1 quart of water. How much did he have left in the bucket. (1 gallon = 4 quarts)

5. He carried the bucket from the kitchen sink to the smoky wall. First, he carried it 9 feet and 8 inches and had to set it down to rest for a moment. Then he carried it 7 feet and 6 inches more to get to the wall. How far is it from the sink to the wall?

.......COMPLETE SOLUTIONS.......

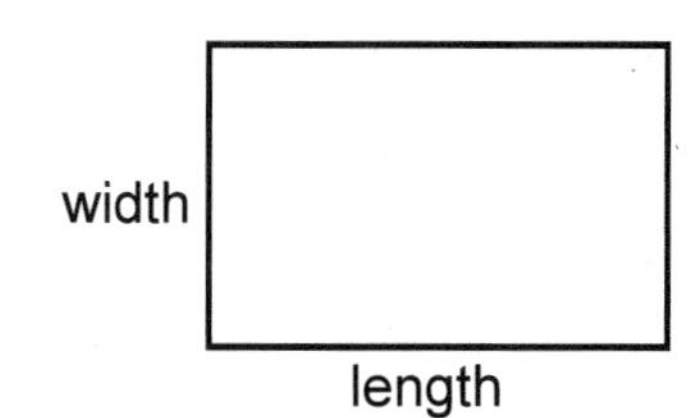

1. The area of a rectangle is length times width. In algebra the formula is $A = \ell w$.

The area of the smoky wall is 8×11, which is 88 square feet.

(In this case, the width of the rectangle is the height of the wall.)

2. The perimeter of a rectangle is the distance around the outside, which is width + length + width + length. In algebra the formula is $P = 2\ell + 2w$.

The perimeter of the wall is $8 + 11 + 8 + 11$, which is 38 feet.

3. The amount of work Fred will have to do depends on the area of the smoky wall. Two walls can have the same perimeter and very different areas.

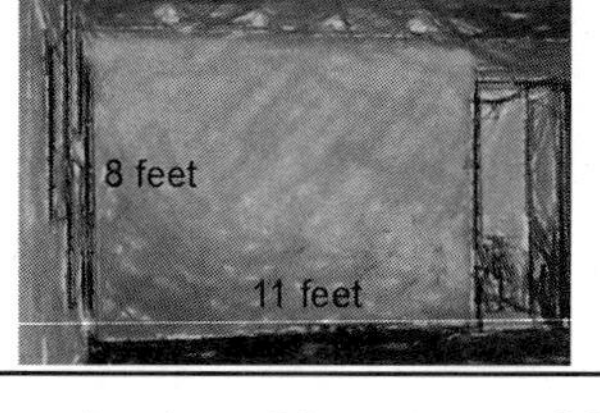

perimeter = 38 area = 88

perimeter = 38 area = 48

4.

4 gals	0 qts		3 gals	4 qts
– 2 gals	1 qt	becomes	2 gals	1 qt
			1 gal	3 qts

He had one gallon and three quarts in his bucket.

5.

```
   9'  8"
 + 7'  6"
  16' 14"  =  16'  12" + 2"  =  16' + 1'  2"  =  17'  2"
```

' means feet and " means inches.

Chapter Fourteen
Daydreaming

With a bucket of warm soapy water and a sponge, Fred scrubbed a circle with a diameter of two feet. He looked at his work. That's a good start he thought. It's nice and round like a circle should be.

Fred started daydreaming about how he taught about circles in his classes at KITTENS.

Fred's Daydream

He started by showing what a diameter is.

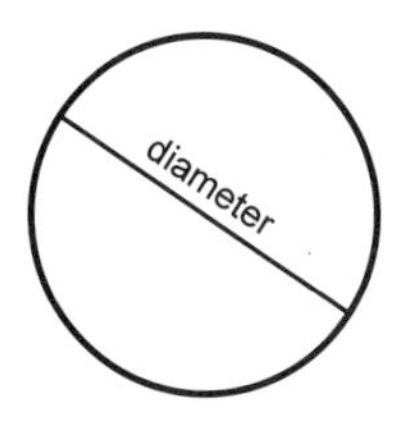

One student, Darlene, called the diameter "the distance across."

Another student, Joe, called it "how fat the circle was."

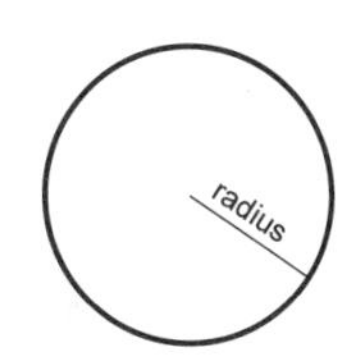

Then Fred talked about the radius of a circle. Darlene said, "The radius starts from the center." She was right.

Fred asked the class, "If the diameter of a circle is two feet, what is the length of the radius?"

Every hand in the class went up, except Joe's. He was busy opening a bag of jelly beans. Everyone else knew that if the diameter was two feet, the radius was one foot.

Fred asked a harder question, "What if the diameter was 72 feet?"

A little short division gave the answer. $\begin{array}{r} 36 \\ 2\overline{)72} \end{array}$

If the diameter is 72, then the radius is 36.

Then he showed what a **chord** of a circle is. He would give lots of examples.

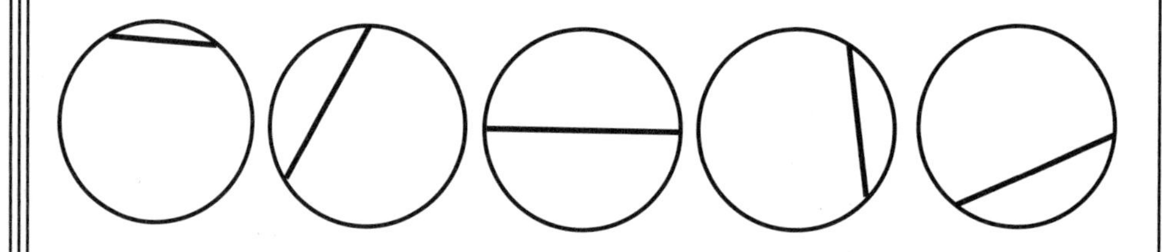

Darlene said, "A chord connects any two points on a circle."

Joe drew on his paper.

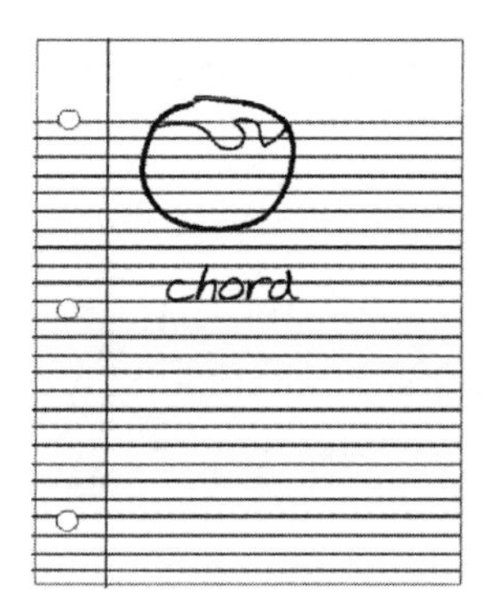

Darlene looked at Joe's notes and yelled at him, "It's gotta be straight!"

Joe corrected his paper.

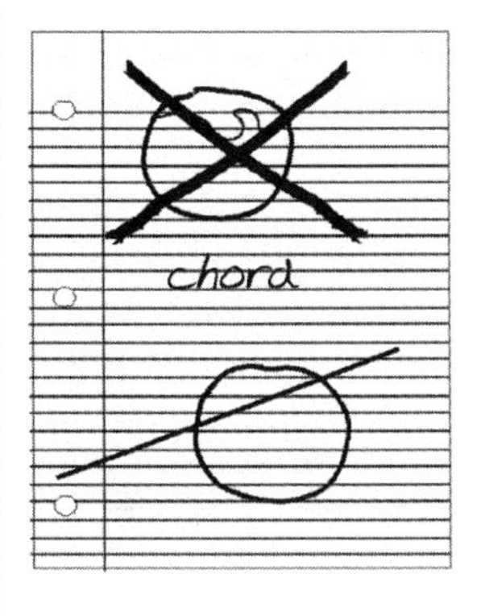

Darlene hit Joe on the arm and shouted, "The chord has to have its ends on the circle." Joe rubbed his arm. Half of his jelly beans had spilled on the floor.

Fred was a much more gentle teacher than Darlene.

Fred asked a harder question, "Suppose you knew what a chord was. How would you define a diameter?"

Joe was still working on what a chord was, so he didn't try to answer this question.

None of the students could think of an answer. Fred had two thoughts:

(1) A diameter is a chord that passes through the center of the circle.

(2) A diameter is the largest possible chord that a circle can have.

Joe swallowed a mouthful of jellybeans and put up his hand. "I got a third definition of a diameter."

Everyone except Fred groaned. Fred was a very patient teacher.

Joe suggested, "A diameter is any chord that contains a radius." He went to the blackboard and drew some diagrams.

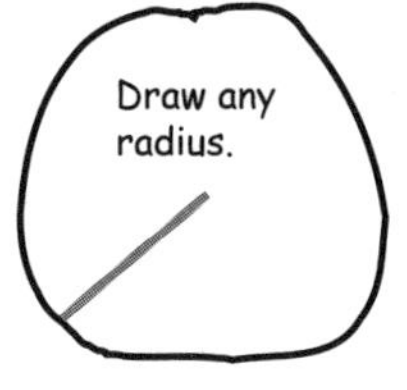

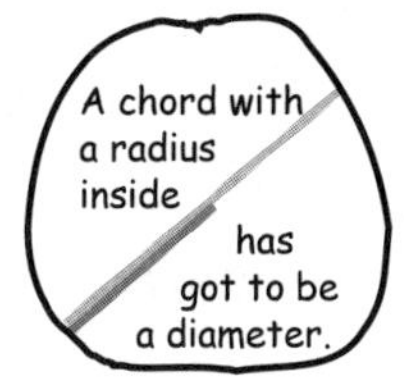

Everyone started to laugh at Joe. They knew he was going to be wrong. He always was.

But everybody was wrong.

The laughter died away.

Darlene said to Joe, "Good job. You really do know what a chord is."

F Major chord

Joe took out another sheet of paper and showed Darlene a chord.

When Fred had talked about chords in circles, at first Joe thought Fred was talking about music.

On his paper, Joe had drawn three circles (whole notes) and that made a chord.

The word *chord* has two different meanings.

There are good things about daydreaming. When you stop doing things and just think about the past or the future, sometimes those moments become some of the best times in life.

★ If you think about some wonderful time you spent last summer . . .

★ If you dream about what you will be in the years to come . . .

Of course, there are also drawbacks to daydreaming. If you are washing the smoke stains off of a wall and you daydream about teaching about circles, you might step into the bucket of soapy water.

Your Turn to Play

1. Joe likes to daydream about food. He dreamed of a pizza that had a diameter of 38 feet.

   What is the radius of that pizza?

2. Joe dreamed of owning his own radio station. He could play his guitar and sing all day long. (*Guitar* has a silent *u* in it.)

   If he owned the radio station, no one could tell him to stop singing. He could even make up songs and sing them.

   If the sounds of his music could be heard for 50 miles in all directions, what is the diameter of that circle?

3. ← is a chord with three circles (whole notes). Draw a circle with three chords in it—not the music chords, but Fred-type chords.

4. In May, Fred had 70 daydreams. Eighty percent (80%) of them were about teaching math. How many math-teaching daydreams did he have?

5. How many were *not* about teaching math?

.......COMPLETE SOLUTIONS.......

1. If the diameter is 38 feet, then the radius is 19 feet.

$$2\overline{)\,3^{1}8\,}^{\;19}$$ using short division

2. If Joe's singing could be heard for 50 miles in all directions, then the radius of the circle is 50 miles.

The diameter would be 100 miles.

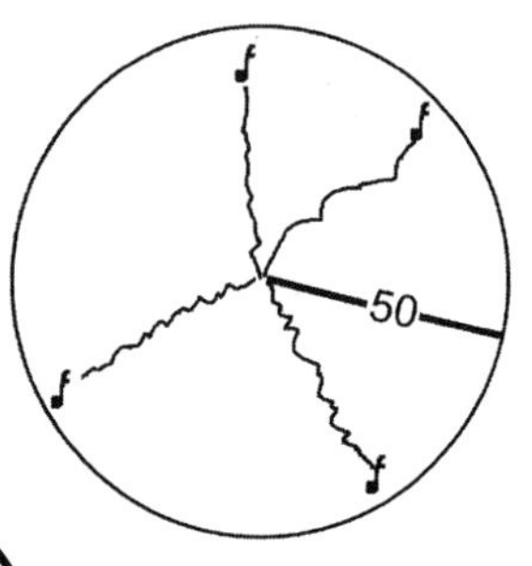

3. Here is a circle with three chords in which none of them intersect.

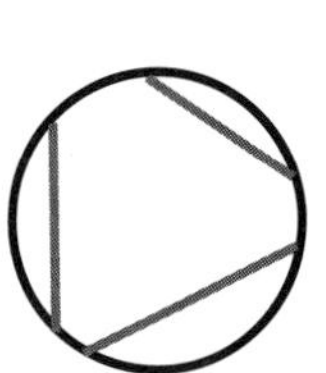

Here is a circle with three chords in which two of them intersect.

4. 80% of 70

$\frac{80}{100}$ of 70

$\frac{4}{5}$ of 70 reducing the fraction by dividing top and bottom by 20*

70 multiplied by 4 and divided by 5 is 56. $70 \times 4 = 280$ $\quad 280 \div 5 = 56$

5. If 56 of the 70 daydreams are related to math teaching, then 14 of them are not. 70 – 56 = 14

(You could have also answered the question: 20%.)

---

* There are several ways to reduce $\frac{80}{100}$

For example, you could first divide by 2 and get $\frac{40}{50}$

and then divide by 2 again and get $\frac{20}{25}$

and then divide by 5 and get $\frac{4}{5}$

Chapter Fifteen
The Parents Arrive

Fred pulled his left foot out of the bucket of soapy water. He walked across the room so that he could get a better view of the wall he was washing. He did not notice two things. First, the carpet had as much smoke damage as the wall. Second, his wet left shoe was cleaning the carpet as he walked.

Shirley appeared in the doorway. She had not gone to sleep like her brothers.

Fred sat down and took off his shoes. He didn't want to make any more marks on the carpet.

Shirley walked over to the bucket and tipped it over. That was fun. She was imagining that this would help clean the carpet.

Fred didn't know how to get the pool of water off the carpet.

It was 10 o'clock. Mr. and Mrs. Abend arrived, and Miss Ente was there to meet them.

Mrs. Abend asked, "And how are my three children? Have they been well-behaved?"

Miss Ente smiled (as much as a horse can smile) and said, "They have been perfect darlings." That might not have been the complete truth since Miss Ente had not seen them since she turned them over to Fred's care several hours earlier.

Shirley came outside and ran to her dad. She said, "The boy, he read me the banananananananananananananananana book."

Mr. Abend picked up his daughter and commented, "Your feet are all wet."

Miss Ente explained that the children were being taken care of by a very short man named Fred. She thought that sounded better than saying that he was a boy. When Fred appeared at the bunkhouse doorway in his bare feet, the Abends knew that they were looking at a boy.

Mrs. Abend went inside to fetch her Gomer and Harry. The room smelled of smoke. The walls and floor were wet. She looked at Fred and asked, "Was there a fire here?"

Fred, who always tells the truth, said, "No. The dinner I was cooking for Harry got burned. It made a lot of smoke."

"You were cooking him dinner? Harry usually eats from the baby food jars that I left here."

"I couldn't get those jars open."

"What in the world were you cooking for my one-year-old?"

Fred's answer was distressing to Mrs. Abend. He said, "Steak."

When she saw Dumma lying in bed and covered with mud, that was the last straw.*

She picked up her two boys and marched out to her husband. "This is the last time we will leave our children at this camp!"

As they drove off, Harry awoke and cried, "I hungry."

Miss Ente saw the bunkhouse plant in the garbage. She went to the bunkhouse and looked inside.

In the history of Camp Horsey-Ducky, there is only one child who has been expelled. Who was this five-year-old who was kicked out?

Fred.

Miss Ente pointed a hoof toward the door and said, "Out." Fred carried his twelve boxes outside into the night.

Except for one light on the outside of the bunkhouse, it was dark. And quiet. And Fred felt sad. He had made more mistakes than he could count. At least, that is how it seemed to him.

---

* When there are a bunch of troubles or irritations, *the last straw* is the one that makes you go nuts.

a trouble ➠ Her daughter's feet were wet.
a trouble ➠ The interior of the bunkhouse was a mess.
a trouble ➠ Some barefoot boy was taking care of her kids.
a trouble ➠ The boy was trying to feed her little Harry a steak.
the last straw ➠ The other camper wasn't cleaned up before he was put to bed.

Camels are used in some desert countries to carry heavy loads. There is an old story about someone who put lots of weight on a camel's back and then added some more weight, and then finally added one little straw. It was that last little straw that broke the camel's back.

It was from that story that *the last straw* entered the English language.

Intermission

There are about seven billion people on earth right now. That is 7 followed by nine zeros. 7,000,000,000

Roughly, six billion of them are at least five years old. 6,000,000,000

And every single one of those six billion has made mistakes that they are sorry about.

All of us. Every adult carries around sadness over mistakes that were made *years* ago. It is just part of being human.

If you are human, you will make errors.

That's why hugs and forgiveness are so important.

Fred arranged the twelve boxes into four stacks of three boxes. His box of six iron frying pans now had only five.

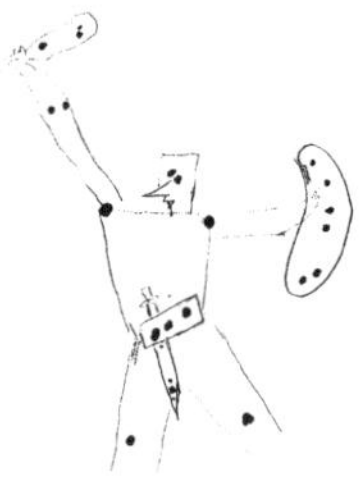

He looked up at the sky and couldn't see the constellation Orion. Maybe, it was because he was crying.

Orion

Miss Ente had rejected him. Mr. and Mrs. Abend did not seem to like him. Only God offered him hugs right now.

Fred looked down the highway that led from Camp Horsey-Ducky to the bus station near the blood bank.

In the daytime that highway looked like this. 

At night it looked like this. 

Your Turn to Play

1. Fred looked through the 80 items in his boxes. Right now he didn't need the sun screen, the pancake turner, or the sundial.

He estimated that right now he needed about five percent (5%) of the items he had packed. How many is that?

2. It was now about 10:15 p.m. Daylight would come about 6 a.m. How long would Fred be in the dark?

3. It was 9 miles from the blood bank to the front door of the bunkhouse. Fred was now 40 feet closer to the blood bank. How far is it from Fred to the blood bank? (one mile = 5,280 feet)

9 miles

.......COMPLETE SOLUTIONS.......

1. Five percent of 80

5% of 80

$\frac{5}{100}$ of 80

80 multiplied by 5 and divided by 100

$$\begin{array}{r} 80 \\ \times\ 5 \\ \hline 400 \end{array} \qquad \begin{array}{r} 4 \\ 100\overline{)400} \\ \underline{400} \end{array}$$

Five percent of 80 is 4.

2. 10:15 to 11:00 is 45 minutes.
11:00 to midnight is 1 hour.
midnight to 6 a.m. is 6 hours.

From 10:15 p.m. to 6 a.m. is 7 hours and 45 minutes.

3.

9 miles			8 miles	5280 feet
−	40 feet	−		40 feet
			8 miles	5240 feet

You may have noticed a pattern for dividing by 100.

$400 \div 100 = 4$

$700 \div 100 = 7$

$8{,}000 \div 100 = 80$

$37{,}000 \div 100 = 370$

$50{,}200 \div 100 = 502$

Chapter Sixteen
Camping

Fred sat on one of the stacks of boxes. He could not carry all twelve boxes for eight miles and 5,240 feet to get to the bus stop near the blood bank. He did not want to go walking in the dark and leave all his stuff behind.

In those boxes were four items that would help him get through the night: an extra-small cowboy hat, neckerchief, rope for cows, gloves to avoid rope burns, silver spurs (with gold trim), **harmonica**, sundial (in case there were no clocks at the camp), mosquito spray, sun screen, campfire songbook, kite with string and extra tails, compass, **a tent**, bandages, ax, canteen, **a sleeping bag**, poison oak soap, **lantern**, pancake turner (for cooking breakfast in the great outdoors), five iron frying pans of various sizes, and a case of flares (to signal for help in case of emergency).

He started with the lantern. It was electric so he didn't need matches. Now he could see more easily.

Next, he set up his tent.*

He rolled out his sleeping bag inside the tent.

Finally, he sat at the opening to his tent with his harmonica and played some camping songs. He was starting to feel better.

He broke into song, "♪♫ Oh, bury my liver out on the lone prairie." He couldn't exactly remember all the words of the original song.

He tried, "Oh, bury my heart in Tennessee." That didn't sound right either.

---

* In official camping language, he pitched his tent.

furry lark named Ben

"Oh, furry the lark Ben misery." That wasn't right either. He couldn't imagine a lark named Ben that was furry.

Dumma came and sat down next to Fred. "I heard the singing. It woke me up. I want to play too."

Fred was happy to have some company. They pretended that the lantern was a campfire and sang songs together.

The electric lantern did not give out much heat. It wasn't a real campfire, and it was getting a little cold. Fred wasn't wearing shoes. His thermometer read 59°.*

After they sang the tenth verse of "Oh hurry my park on the roan** Marie," Dumma was feeling tired again. It was well past his normal bedtime. Without asking permission, he crawled into Fred's tent and lay down on top of Fred's sleeping bag. He was too tall to get into Fred's little sleeping bag.

---

* It was 59°F (where F stands for Fahrenheit). Much of the world outside of the United States uses the Celsius scale, which is in the metric system.

To change from Fahrenheit to Celsius you do two things: First, you subtract 32. Second, you multiply by $\frac{5}{9}$ (which means multiply by 5 and divide by 9).

To change 59°F to Celsius, you first subtract 32°. $59 - 32 = 27$

Then you multiply by $\frac{5}{9}$ which means 27 times 5 and divide by 9.

$$\begin{array}{r} 27 \\ \times\ 5 \\ \hline 135 \end{array} \qquad 9\overline{)1\,3^{4}5}^{\;1\,5} \text{ using short division}$$

59°F = 15°C

** *Roan* is a color of horses. A roan horse is reddish brown or black speckled with white or gray.

Dumma was 80% of Fred's age, but he was 125% of Fred's height.

Fred was 5 years old.

80% of 5 years

$\frac{80}{100}$ of 5

$\frac{4}{5}$ of 5 reducing the fraction by dividing top and bottom by 20 (or, if you wish, by dividing by 5 and then by 4).

5 multiplied by 4 and divided by 5

$5 \times 4 = 20$

$20 \div 5 = 4$

Dumma is 4 years old.

Fred was 36 inches tall.

125% of 36 inches

$\frac{125}{100}$ of 36

$\frac{5}{4}$ of 36 reducing the fraction by dividing top and bottom by 25 (or, if you wish, by dividing by 5 and then dividing by 5 again).

36 multiplied by 5 and divided by 4

$$\begin{array}{r} 36 \\ \times\ 5 \\ \hline 180 \end{array} \qquad 4\overline{)180} = 45$$

Dumma is 45 inches tall.

It was suddenly quiet. Miss Ente was grateful. The boys were camping in the parking lot of Camp Horsey-Ducky, and their singing had been keeping her awake.

Fred looked in the tent. Dumma was asleep. Fred's sleeping bag was now very dirty. Fred thought When I grow up and I get married and have a son and he has been playing in the mud . . .

I will remember to wash him before he goes to bed.

The tent was small. Dumma was much bigger than Fred. He took up the whole space inside the tent.

Fred headed back to the twelve boxes and found the campfire songbook. He found out that the song he had been trying to sing was "Bury Me Not on the Lone Prairie."

He looked at the thermometer again. It read 50°. In his head, he converted 50°F to Celsius. First, you subtract 32 and get 18. Then you multiply by $\frac{5}{9}$ so 18 times 5 is 90 and divide by 9 is 10. Fred was good at mental arithmetic. Most five-year-olds could not convert 50°F to 10°C even if they were allowed to use pencil and paper.

Fred looked through the boxes and put on every shirt and coat that he could find. That made him look very round.

He started to feel very hot and very cold at the same time.

He took off some of the coats and put on some socks and shoes. He no longer looked like a barefoot Santa Claus.

Fred invented the Comfortableness function. The domain would be a set of situations and the codomain would be {comfortable, not comfortable}.

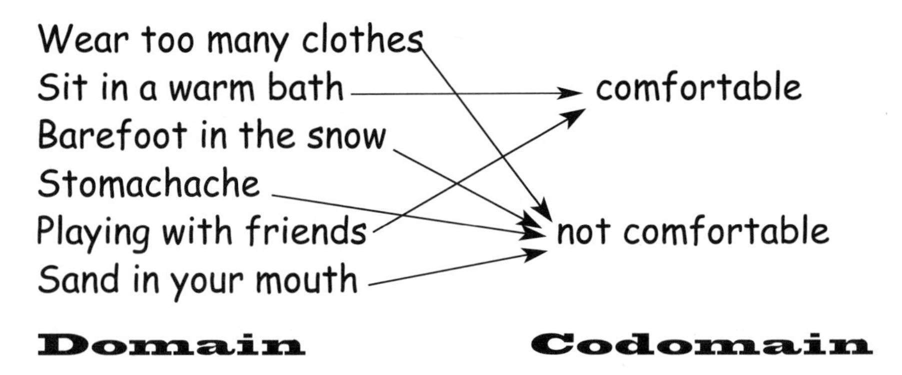

It is a function because each element of the domain has exactly one image in the codomain.

Your Turn to Play

1. Soon it was going to be 41°F. Convert 41°F to Celsius.

2. Water boils at 212°F. How much is that on the Celsius scale?

3. Water freezes at 32°F. Convert that to Celsius.

4. What is the cardinality of {Fred, Dumma, Shirley, Gomer, Harry}? Recall: The cardinality of {#, ☢} is 2.

5. The plural of *diameter* is *diameters.*
The plural of *radius* is *radii*. (RAY-dee-eye)
In a given circle are the radii all equal in length?

6. In a given circle are all chords equal in length?

7. You have a circle with diameter equal to 10 inches.
A) Is it possible to have a chord equal to 2 inches?
B) Is it possible to have a chord equal to 15 inches?
C) Is every chord that is equal to 10 inches a diameter?
D) (harder question) Can a chord ever be a radius?

.......COMPLETE SOLUTIONS.......

1. To convert from Fahrenheit to Celsius, you subtract 32 and multiply by $\frac{5}{9}$. $41 - 32 = 9$

To multiply 9 by $\frac{5}{9}$ you multiply by 5 and divide the result by 9.

$9 \times 5 = 45$ $45 \div 9 = 5$. $41°F = 5°C$

2. $212 - 32 = 180$

180 times 5 and divide by 9

$180 \times 5 = 900$ $900 \div 9 = 100$

$212°F = 100°C$

3. $32 - 32 = 0$

0 times 5 and divide by 9

$0 \times 5 = 0$ $0 \div 9 = 0$

$32°F = 0°C$

So in the metric system, water boils at 100°C and freezes at 0°C. That is much easier to remember than 212°F and 32°F.

4. The cardinality of {Fred, Dumma, Shirley, Gomer, Harry} is 5.

5. Yes. If, for example, a circle had a diameter of 18, then all the radii would equal 9.

6. No.

7A. Yes. You can have a chord as close to zero inches as you like.

7B. No. It would not fit inside the circle.

7C. Yes. Every 10-inch chord would have to pass through the center of the circle.

7D. No radius is a chord. One end of every radius is at the center of the circle.

radius

Chapter Seventeen
Into the Night

Fred had been kicked out of Camp Horsey-Ducky. He was camping in the parking lot. His tent was occupied by Dumma. He couldn't get to the bus stop at night with all of his stuff. He couldn't sing any more campfire songs because that might wake up Dumma.

In short, he didn't know what to do with himself.

small essay

Questions at Each Stage of Life

Babies never ask, "What is there to do?" They just hang around. They give very little thought to the next 70 or 80 years of life.

Kids ask their moms, "What shall I do? I'm bored."

Teenagers ask a million questions as they seek to find out who they are, where they are headed, and with whom they will be going there.

Young adults in the first half of life ask, "Where can I find enough time to do all the stuff that needs to be done?" There is work to do, clothes to wash, taxes to pay, etc.

At the halfway point, say around 40, some may ask, "Is this all there is? Am I missing something? Do I need to change the direction of my life?"

In their sixties, some will think, "No job to go to. The kids are all grown and out of the house. Do I putter in the garden? What kind of vacation can we afford?"

In the last years of life, "I wonder what Heaven will really be like? I'm looking forward to seeing Mom and Dad and old friends."

end of small essay

At this point Fred wasn't very sleepy. He picked up the lantern and decided to go exploring around Camp Horsey-Ducky. After all he thought what could be wrong with walking around a little bit at night?

Sometimes five-year-olds do not make wise decisions.

He didn't walk toward the bunkhouse. He didn't walk toward the highway. He headed perpendicularly to the line that connected those two.

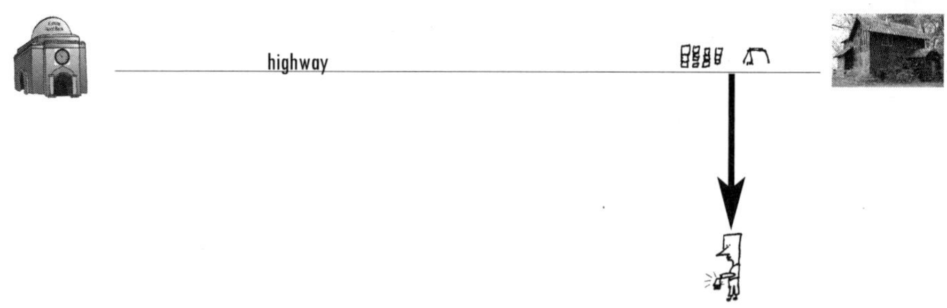

He had heard Miss Ente talking with the men from the Friendly Fence Company just after Methhunda had left. They were talking about installing an eight-foot fence around the perimeter of Camp Horsey-Ducky.

✓ Fred knew that if he came to an eight-foot fence, that would be the border of the property.

✓ He knew that if he stayed on the property he would be safe.

> Note to reader: Would you care to guess how many of these things that Fred knew were correct?

It had been half past six in the evening when the Ganses and Methhunda had arrived at the camp. It was after that when the fence workers arrived. They probably did not install the 636 feet of fencing that night.

Fred walked past the four rose gardens. He was glad that he had a lantern. Otherwise, he might have accidently stumbled into a rose bush and gotten scratched.

He kept walking. The light from the lantern dimmed as the batteries began to wear out. Soon, the only light Fred had was from a crescent moon.

He walked slowly for about a half an hour and began to wonder when he would come to the fence that marked the boundary of Camp Horsey-Ducky.

He did not realize that twenty minutes ago he had left the camp property.

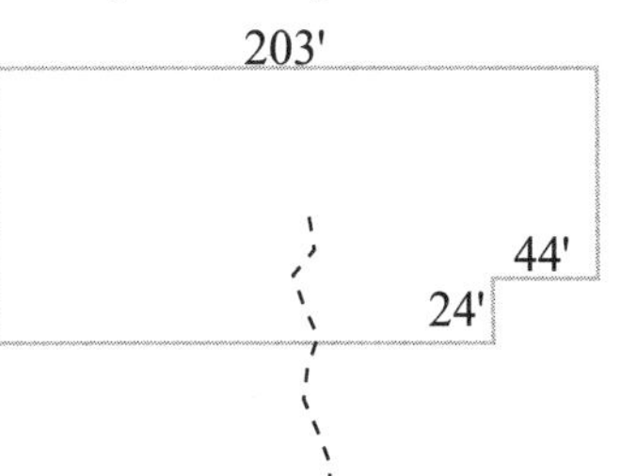

What was out there on the flat plains of Kansas in the darkness?

Was there a **m**oose armed with a **m**achine gun?

No. There is not a single **m**oose in Kansas that carries a weapon.

Was there a 500-pound **m**osquito ready to attack Fred?

No. No **m**osquito in Kansas weighs more than 200 pounds.

Was it **M**oby Dick, the great white whale?

No. Don't be silly. **M**oby Dick lives in the ocean and Kansas is about as far away from an ocean as you can get.

Wait a minute! I, your reader, have a question. All these suggestions you have been making start with the letter m: moose, machine gun, mosquito, Moby Dick. Are you hinting that what Fred is going to encounter starts with the letter m?

Yes.

In front of Fred was a wire fence. The wire was one meter off the ground.

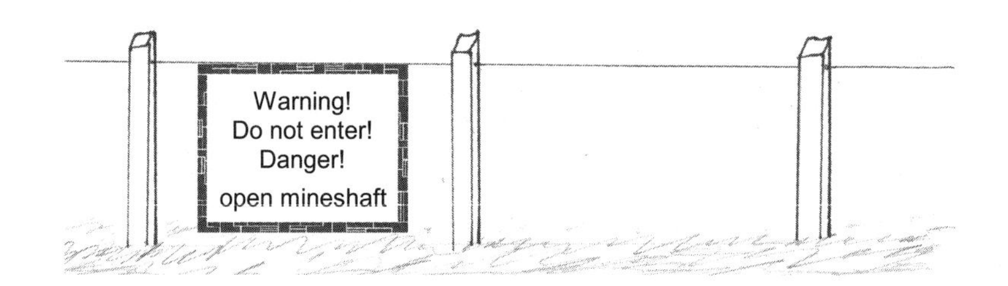

Many abandoned **m**ines have no fencing to warn people. This hole in the ground was so dangerous that someone put fence posts in the ground and strung a wire at a height that would keep horses and adults from accidentally stumbling into the hole.*

---

* The land around KITTENS University in Kansas is pretty flat. If you want to dig a mine, you have to dig straight down and create a mineshaft. In the movies, whenever you see a mine, it is always dug into the side of a mountain, and people walk into the mine.

Your Turn to Play

Example: There are 16 ounces in a pound. If you wanted to convert 50 ounces into pounds, you divide 16 into 50.

$$16\overline{)\,50\,}\quad 3\ R\ 2$$

$$\begin{array}{r} 3\ R\ 2 \\ 16\,\overline{)\;50} \\ \underline{48} \\ 2 \end{array}$$

50 ounces = 3 pounds and 2 ounces

Example: There are 7 days in a week. Three hundred days is equal to how many weeks?

$$\begin{array}{r} 42\ \ R\ 6 \\ 7\,\overline{)\;300} \end{array}$$

300 days = 42 weeks and 6 days

1. Here is where mathematics becomes very important in Fred's story. The fence wire is one meter off the ground. Fred is 3 feet tall. One meter is equal to about 39 inches.

   Convert 39 inches into feet. (There are 12 inches in a foot.)

$<$ means "less than"

2. Which is correct:

   height of the wire $<$ Fred's height or

   Fred's height $<$ height of the wire.

3. Suppose we have two numbers. Call them x and y. (We do this all the time in algebra.)

   Suppose we know that $x < y$.

   Must it be true that $y > x$?

.......COMPLETE SOLUTIONS.......

1. To convert 39 inches into feet . . .

$$\begin{array}{r} 3 \text{ R } 3 \\ 12\overline{)\,39} \\ \underline{36} \\ 3 \end{array}$$

39 inches = 3 feet and 3 inches

So one meter = 3 feet and 3 inches

2. Fred's height is 36 inches. The height of the wire is 39 inches. Therefore, Fred's height < height of the wire.

3. If x is less than y, then it must be true that y is greater than x. If 5 is less than 8, then 8 must be greater than 5.

If 30 is less than 97, then 97 must be greater than 30.

If $x < y$, then it must be true that $y > x$.

A meter isn't exactly 39 inches. It would be more accurate to say that a meter is about 39⅓ inches.

The only decimals that you have probably seen at this point in your life are the decimals in money. $7.29 means seven dollars and 29 cents.

After *Life of Fred: Fractions* comes *Life of Fred: Decimals and Percents.* At that point, we could say that one meter is about 39.370078740157 inches, but it would be silly to mention that now.

Chapter Eighteen
Mineshaft

Fred did not see the warning sign. It was too dark. He did not see the fence wire. Since 36" < 39", he just walked under the wire not knowing it was there. Exploring abandoned mines can be dangerous. Falling into mineshafts is worse.[*]

The mineshaft
near
Camp Horsey-Ducky

"Hey, son! Don't take another step. There's a mineshaft right in front of you."

Fred had thought he was alone.

"Turn around and head back to your tent."

That was a deep voice that Fred had never heard before.

---

[*] When my younger daughter was three years old, she once mentioned that when you do something dangerous "you could get killed to death." When she went to college, she decided to be an English major. I wasn't surprised.

Fred turned around and headed back toward his tent. He had been three steps away from death.

Fred thought about that voice for the rest of his life.

When he was thirteen, he came back to that mineshaft during the daytime. He looked down into the black hole. If he had fallen into the hole, he would have been killed, and it might have been days or weeks until someone found his body.

In contrast, if Shirley were missing, Mr. and Mrs. Abend would be searching for her immediately.

It wasn't until years later he heard a song with the words, "He orders his angels to protect you, to keep you from smashing against the rocks." Then it all made sense.

Fred walked back to his campsite. Dumma was sleeping soundly on Fred's sleeping bag. It was nearly midnight and Fred needed to find a place to sleep.

He opened up one of the boxes and looked through it hoping to find an idea of how he could get to sleep. He took out the silver spurs, the sundial, and the kite with string and extra tails. None of those will help he thought to himself.

He couldn't reach the items at the bottom of the box so he crawled into the box. He found some batteries and put them into his lantern.

Much better. In the bottom of the box he found an ad from King KITTENS, the largest store in town. Great! This can be my bedtime reading.

Midnight Sale

Your very own sandbox!
Four seats!
Super shade!
Normally $435.
Now 40% off!
Sand not included.

Fred did the computation in his head . . .

40% of $435

$\frac{40}{100}$ of 435

$\frac{2}{5}$ of 435

He reduced the fraction by dividing top and bottom by 20. It could have been done by dividing by 10 and then by 2.

$\frac{2}{5}$ of 435 means 435 times 2 and divide by 5.

$$\begin{array}{r} 435 \\ \times\quad 2 \\ \hline 870 \end{array} \qquad\qquad \begin{array}{r} 174 \\ 5\overline{)\,8^{3}7^{2}0} \end{array}$$

40% of $435 is $174.
He could save $174.

That means that the sale price would be $261.

$$\begin{array}{lr} \text{regular price} & 435 \\ \text{savings} & -\,174 \\ \cline{2-2} \text{sale price} & 261 \end{array}$$

All kinds of thoughts ran through Fred's head . . .

1. I could be the only faculty member at KITTENS that has a sandbox.
2. Where could I put it? My office is too small. Maybe it could go out in the hallway.
3. If I put it in the hallway, then other people could also use it.
4. I wonder how much the sand costs?
5. I wonder what it would be like to put my sleeping bag on the sand and sleep out in the hallway? Would that be better than sleeping under my desk? Sleeping is fun. Sleep.

Fred put his head down and fell asleep in the box.

He never noticed that the ad was for January 20, 1962.

Your Turn to Play

1. (not easy) A bag of sand is 27 pounds. The sandbox needs at least 115 pounds of sand. How many bags would Fred need to get?

2. A bag of sand normally costs ninety-five cents (95¢). For this sale, it was 40% off. How much is the sale price for one bag?

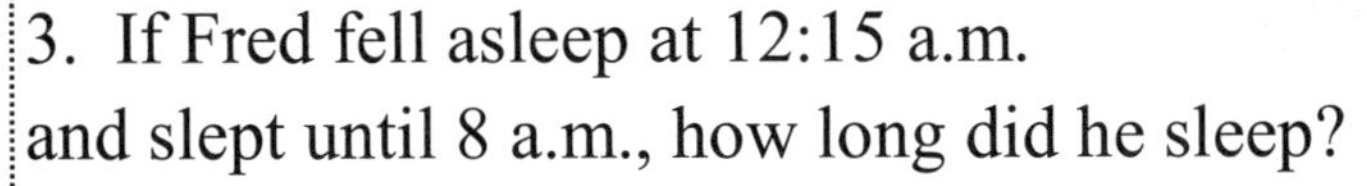

3. If Fred fell asleep at 12:15 a.m. and slept until 8 a.m., how long did he sleep?

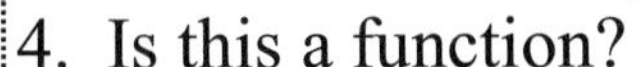

4. Is this a function?

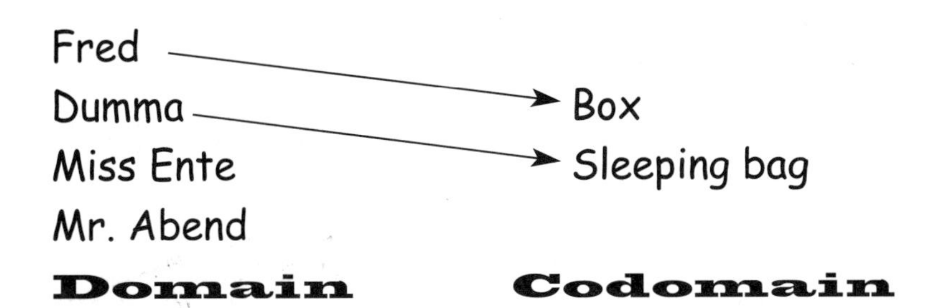

5. Is this a function?

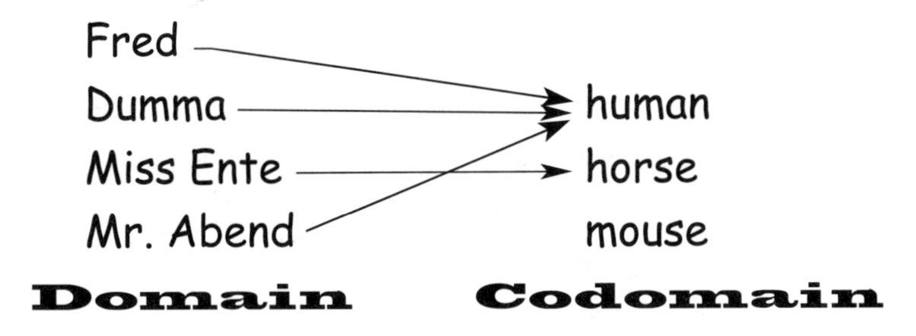

.......COMPLETE SOLUTIONS.......

1. The first thing you would normally do to solve this problem is divide 27 into 115. Using the general rule, you might have restated the problem with a bag of sand weighing 3 pounds and you needed 12 pounds of sand. Then you would have noted that it was a division problem.

But in this case, when you divided you would have:

$$\begin{array}{r} 4 \\ 27\overline{)115} \\ \underline{108} \\ 7 \end{array}$$

It didn't go in evenly.

Four bags would not be enough since $4 \times 27 = 108$ pounds.

Five bags would give more than 115 pounds since $5 \times 27 = 135$ pounds.

Carefully reading the problem, you will see that it asks for *at least* 115 pounds. So Fred would need 5 bags.

2. 40% of 95¢

$\frac{40}{100}$ of 95

$\frac{2}{5}$ of 95 which means 95 times 2 and divide by 5

$$\begin{array}{r} 95 \\ \underline{\times\ 2} \\ 190 \end{array} \qquad \begin{array}{r} 38 \\ 5\overline{)190} \end{array}$$

So 40% of 95¢ is 38¢. He saves 38¢.

Regular price minus savings = sale price $95 - 38 = 57$¢

3. From 12:15 to 1:00 is 45 minutes.

From 1:00 to 8:00 is 7 hours.

He slept 7 hours and 45 minutes.

4. It is not a function. Every member of the domain must have exactly one image in the codomain. Miss Ente, for example, does not have any image in the codomain.

5. It is a function. Each element of the domain has exactly one image in the codomain. That's the definition of a function.

Chapter Nineteen
Home

Fred slept for seven hours and 45 minutes. Forty-five minutes is three-fourths ($\frac{3}{4}$) of an hour.

There are two ways to know that 45 minutes is three-fourths of an hour.

One way:

You can count by fives on the face of a clock.

Second way:

Three-fourths of an hour

$\frac{3}{4}$ of an hour

$\frac{3}{4}$ of 60 minutes

60 times 3 and divide by 4

$$\begin{array}{r} 60 \\ \times\ 3 \\ \hline 180 \end{array} \qquad \begin{array}{r} 45 \\ 4\overline{)180} \end{array}$$

Three-fourths of an hour is 45 minutes.

Seven hours and 45 minutes is not enough sleep for a five-year-old. Dumma looked in the box and found Fred. He shouted, "You want to play in the dirt with me?" He knew that he had to shout since Fred was still asleep.

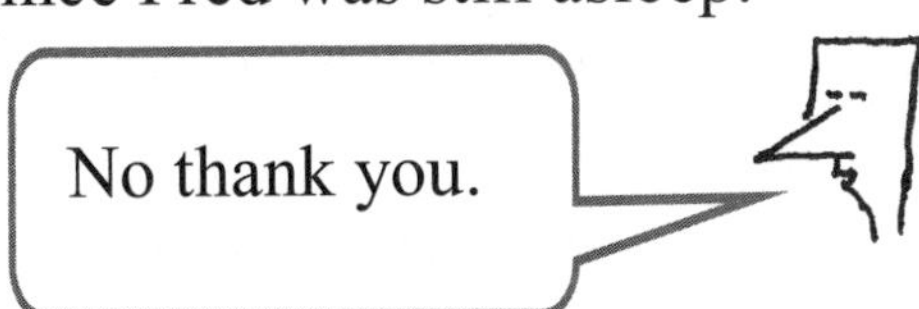

Fred was always polite, even when he was half asleep.

He rubbed his eyes and stretched. He stood up and looked over the edge of the box. A whole new day was beginning.

He hopped out of the box. He tossed the silver spurs, the sundial, the kite with string and extra tails into the box. He pulled the sleeping bag out of the tent and shook it to get rid of some of the dirt. He knew that he would have to wash it when he got back home to his office.

He took down the tent.* He put the tent and the sleeping bag into the box.

Dumma wandered into the bunkhouse looking for some breakfast. He spotted the clean spot on the wall and left his hand print on it. Dumma was four years old. His motto was: Leave no clean spot undirtied.

Fred needed to figure out how to get home. He could walk eight miles and 5,240 feet down the highway to the bus stop, but then he would have to leave his twelve boxes of camping stuff behind.

It was now 68°F.** That's a nice temperature for walking. If you had to memorize just one temperature conversion, the 68°F ↔ 20°C might be the most useful. Both

---

* In official camping language, he struck the tent. One meaning of *strike* is to take down or pull apart.

** To change to Celsius, you subtract 32 $(68 - 32 = 36)$ and then multiply by $\frac{5}{9}$ (36 multiplied by 5 and divided by 9).

$$36 \times 5 = 180 \qquad 180 \div 9 = 20°\text{C}$$

of these are roughly room temperature. If someone says 68°F or says 20°C, think "comfortable."

Every temperature in Fahrenheit has exactly one temperature in Celsius that corresponds to it. That's a function.

68°F is assigned to 20°C.
59°F is assigned to 15°C.
212°F is assigned to 100°C.
32°F is assigned to 0°C.*
etc.

0°C is the **image** of 32°F using this function.

---

* When you get to *Life of Fred: Fractions*, you will be able to do the tricky ones like converting 63°F into Celsius. If we did that now, it would be really confusing.

I, your reader, demand you do it NOW.

Please. Let's wait.

Now!

Okay.

First you subtract 32. 63 – 32 = 31.

I'm not lost yet. Anybody can subtract 32.

Then you multiply by $\frac{5}{9}$ (31 times 5 and divide by 9).

$$\begin{array}{r} 31 \\ \times\ 5 \\ \hline 155 \end{array}$$

Mr. Author, did you think that was hard? You told me this was tough.

And then you divide 155 by 9. $\begin{array}{r} 17\frac{2}{9} \\ 9\overline{)1\,5^{6}5} \end{array}$

So 63°F equals $17\frac{2}{9}$ °C.

What? How did you get the $\frac{2}{9}$? It should have been "17 R 2" (17 with a remainder of 2).

I shouldn't have done this now. I'm sorry. I should have waited until we got to *Life of Fred: Fractions*.

That's okay, Mr. Author. I forgive you.

Fred put on one of his bow ties. That gave him something to do while he was thinking.

He thought of renting a helicopter to come and pick him and his boxes up.

He thought of putting each box on roller skates and pushing them down the highway.

He thought of holding a yard sale and selling all of his stuff. Then he could get a cab and go home.

Then he looked around and saw something that he hadn't seen in the dark. A KITTENS campus mailbox.

Their motto is: Free delivery and faster than email.

He addressed each box: Room 314, Math Building, KITTENS University. He put the first eleven boxes into the mailbox. Then he hopped into the twelfth box as it slid into the mailbox.

The light went on.

One second later, Fred hopped out of the box. He was in his office. Kingie was finishing up another oil painting.

It was good to be home.

Index

If you would like to read more about Fred or order books . . .

FredGauss.com

Polka Dot Publishing

We are proud of our low prices.

Life of Fred: Fractions	$19
Life of Fred: Decimals and Percents	$19
Life of Fred: Elementary Physics	$29
Life of Fred: Pre-Algebra 1 with Biology	$29
Life of Fred: Pre-Algebra 2 with Economics	$29
Life of Fred: Beginning Algebra Expanded Edition	$39
Zillions of Practice Problems: Beginning Algebra (optional)	$29
Life of Fred: Advanced Algebra Expanded Edition	$39
Zillions of Practice Problems: Advanced Algebra (optional)	$29
Life of Fred: Geometry	$39
answer key *Life of Fred: Geometry*	$6
Life of Fred: Trigonometry Expanded Edition	$39
Life of Fred: Calculus — *Two years of college calculus.*	$39
answer key *Life of Fred: Calculus*	$6
Life of Fred: Statistics — *A year of college statistics.*	$39
answer key *Life of Fred: Statistics*	$6
Life of Fred: Linear Algebra	$49
answer key *Life of Fred: Linear Algebra*	$6

Linear algebra is a math course that is required of almost all math majors in college. It is usually taught after calculus.

Order again from:
JOY Center of Learning
http://www.LifeofFredMath.com